Electronics Circuit SPICE Simulations with LTspice

A Schematic Based Approach

Amit Kumar Singh

Rohit Singh

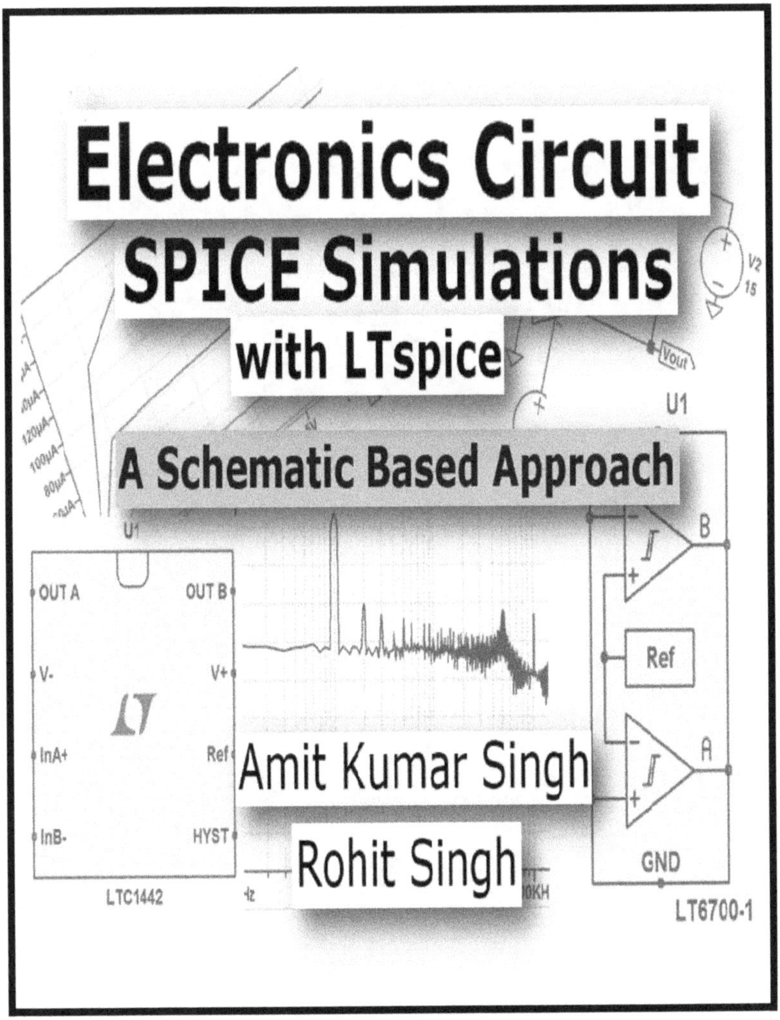
U1
OUT A
OUT B
V-
V+
InA+
Ref
InB-
HYST
LTC1442
Vout
V2
15
U1
B
Ref
A
GND
LT6700-1

Electronics Circuit SPICE Simulations with LTspice

A Schematic Based Approach

First Edition

Amit Kumar Singh
Rohit Singh

createspace Independent Publishing

Copyright © 2015 Amit Kumar Singh & Rohit Singh

ISBN-13: 978-1508649212

ISBN-10: 1508649219

createspace Independent Publishing
www.createspace.com

Electronics Circuit SPICE Simulations with LTspice

A Schematic Based Approach

Contents

Preface to the First Edition

The idea of this book evolved when we started guiding our junior students on SPICE Simulations. By that time we had already used many SPICE simulators. We found LTspice much easier to work with. Also it's free and has a good documentation available online. We felt the need to have a book that could guide the beginners in the field. We have tried to keep things simple and every simulation comes with a background theory. Hope this book is helpful to SPICE Users Community and the students.

We would like to thank Dr. Sonal Singhal, our teacher and guide for her constant support and mentoring. We would like to thank our friends, Jyotirmoy Dutta, Mangal Das, Praveen Dwivedi, Manish Gupta and Saurabh Porwal because from them we have learnt many a things and to our juniors at Shiv Nadar University Arif Khan, Ajay Sharma, Sunil Niranjan, Varun Vaid, Mahesh Vaidya and Namrata Srivastava, we learnt many a concepts with them.

Finally we thank our parents for their immense love and support.

Amit Kumar Singh   Jaipur, 30th July 2015

Rohit Singh

Electronics Circuit SPICE Simulations with LTspice

A Schematic Based Approach

Chapter -1   Introduction to circuit Simulations

Before We Begin

Designing electronics hardware can be a time taking and costly affair. Electronic Circuit Simulators are computer programs that replicate the circuit behavior of Electronic Circuits on a Personal Computer. Electronic Circuit simulators not only give useful insights into the circuit behavior but also help us in minimizing the efforts in terms of time and money. In the design of any electronics circuit, it is important that the simulation of the circuit be done before a hardware implementation is done. One goes for hardware implementation only when accurate simulation results are obtained.

Electronics can be broadly categorized as Analog Electronics and Digital Electronics. Different Electronics Simulators are used both for Analog and Digital Electronic circuits. In Digital Electronics the widely used simulators are based on Verilog and VHDL while in Analog Electronics SPICE based simulators are widely used. In this book we focus on Analog Circuit Simulation using a Simulator based on SPICE.

Introduction

Since this book is all about Spice Circuit Simulations, it is essential that we discuss, what is SPICE? And why do we need it?

Can you imagine food without any Spice? But this SPICE is different. This is the SPICE of electronics. In today's scenario you can't imagine electronics or VLSI industry for that matter without SPICE.

SPICE stands for simulation Program with Integrated Circuit Emphasis. SPICE is a software which takes circuit description as input (which may be provided either in the form of a text file or a schematic file) and it gives output in the form of text or waveforms. SPICE was developed at the University of California at Berkeley in 1975. There are a plethora of software tools available currently in the markets that do SPICE analysis. Some of the available SPICE tools are ORCAD PSpice, NgSpice, HSPICE, XSPICE, TSPICE. Various other SPICE simulators are available in the market. Some of these simulators are freely available for download. In this book, all the simulations will be done on LTspice. Also the Focus will be Schematic based simulation only.

Why LTspice?

LTspice is available free from Linear Technology. LTspice is perhaps one of the most widely used free simulators. It is a powerful simulator with a simple interface to handle. Many other SPICE Simulators are costly and beyond the reach of many students. LTspice IV can be easily downloaded from the Linear Technology's Website, www.linear.com.

Limitations of SPICE

Nevertheless SPICE has certain limitations as well. The accuracy of results obtained by SPICE depends on the MODEL file used in the simulation. Minor errors in simulation results may also result due to the technique adopted by SPICE to solve the equations. SPICE uses Newton-Iteration method to solve the circuit equations.

Notes for the Readers

This book may not be considered as a complete Text. The book intends to present the simulation of simple Analog Electronics Circuits. It may be supplemented with any standard Text. It is also suggested that the reader performs all the simulations on the simulator and verifies the results. Tweaking with the parameters may help one in understanding the circuits in a better manner.

Launching LTSpice

Once you have installed LTSpice in your computer, double click on the LTSpice icon that is available on your desktop or on the drive you have stored its icon.

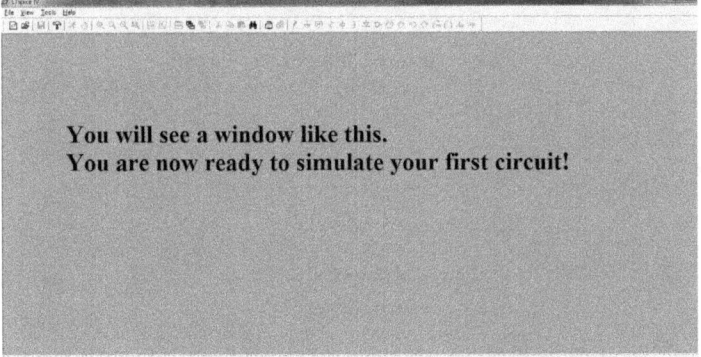

Fig 1.1: LTspice First Look

To start your first circuit simulation, go to the file menu and click on new schematic from your mouse. The schematic icon Looks like this:

The other way is to simply click on the new schematic icon that is available on the extreme left on the toolbar.

Once you do this-

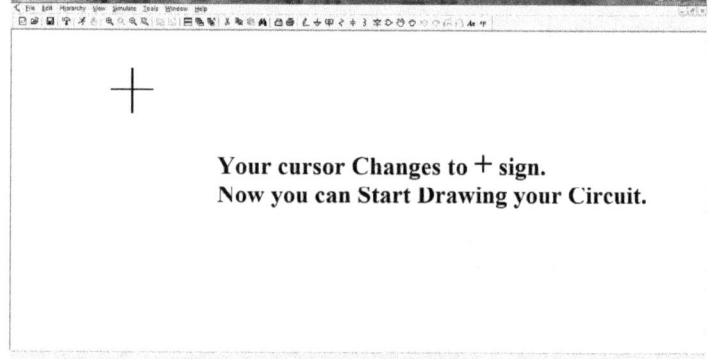

Your cursor Changes to + sign.
Now you can Start Drawing your Circuit.

Fig 1.2: LTSpice Start Look

Now we need to look at the components that we have. If you look closely at the toolbar, you can see resistor, capacitor, diode and component icon present as shown in figure 1.3. When you click on a component it is selected. You can now bring it on the schematic screen.

Click on the component icon D . When you do so, a window appears like as in fig 1.4.

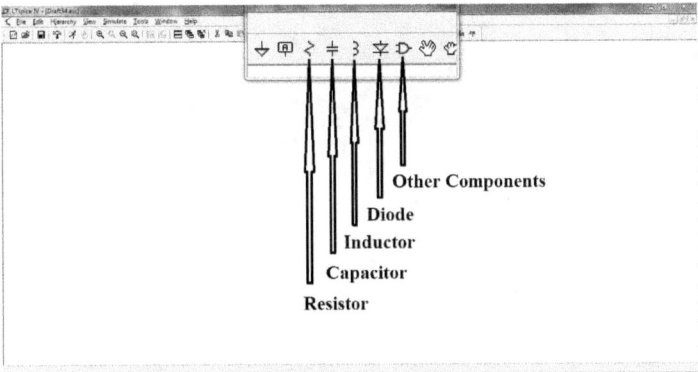

Other Components

Diode

Inductor

Capacitor

Resistor

Fig 1.3: LTSpice components at the toolbar

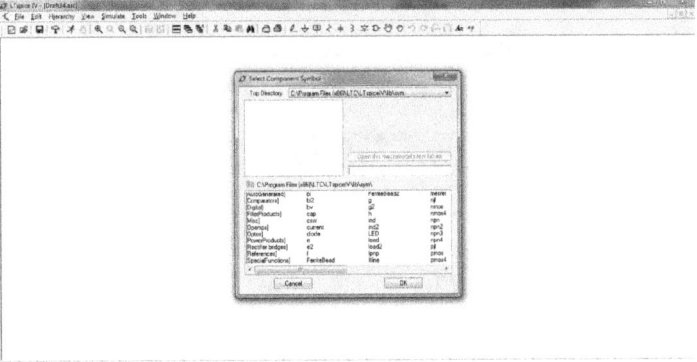

Fig 1.4: Component Selection pop up

In the component pop up you can type the name of the component that you want in your schematic. For example if you want a Voltage source. Type voltage, when Voltage is selected, click OK.

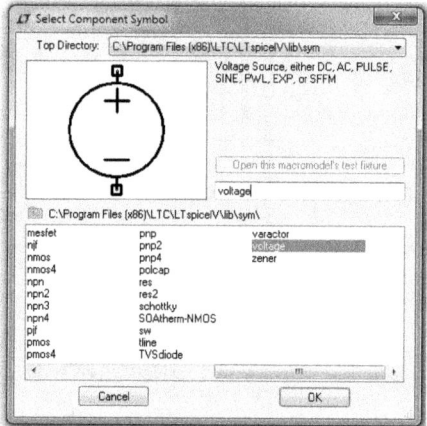

Fig 1.5: Component Selection- Voltage

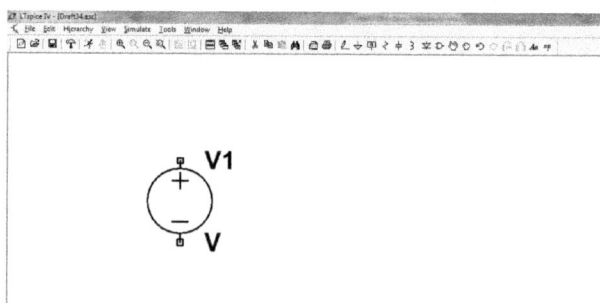

Fig 1.6: Voltage on the schematic

When you place the Voltage source, your Schematic looks something like figure 1.6. We are now ready to simulate our first circuit.

****

Chapter 2   Your First Circuit Simulation

Introduction
In this chapter we will be simulating our first circuit using LTSpice. Note that your Simulator background may look different from that shown in this book. This is because Color Preferences and line width have been modified.

Color preferences can be changed from: Tools-> Color preferences and
Line width from: Tools-> Control Panel -> Drafting Options, as shown in Fig 2.1 and Fig 2.2.

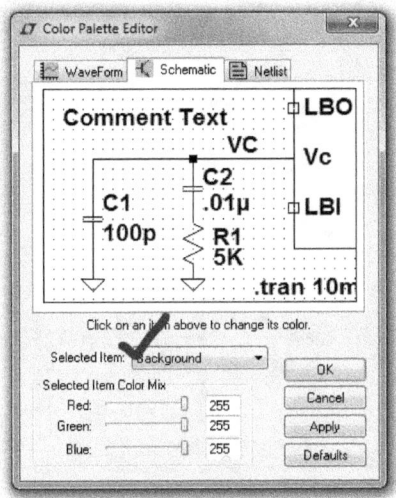

Fig 2.1: Color Palette Editor

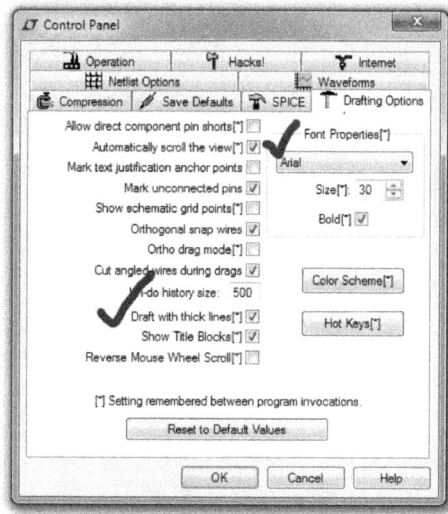

Fig 2.2: Control Panel

Your First Circuit

The aim of this experiment is to verify ohms law, a fundamental law of electricity.
According to Ohm's law, the current through a conductor between two points is
directly proportional to the Voltage across the two points given a constant temperature.
This law can be mathematically presented as

$$I = \frac{V}{R}$$

Or,

$$V = IR \tag{2.1}$$

Equation 2.1 is known as ohms law and is a fundamental law of electricity.

Where I is the current through the conductor in Amperes, V is Voltage across the two points in
Volts, R is the resistance in ohms which is a constant here.

Your First Circuit Simulation

The Circuit Diagram

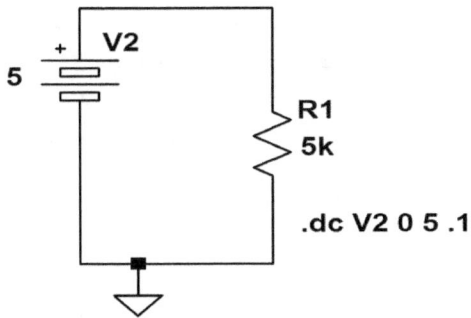

Fig 2.3: A simple schematic

The above circuit contains a battery of 5V named V2 and a resistance named R1 of 5kΩ.The circuit also contains a ground. Every circuit must contain a ground. If a circuit doesn't have a ground, the simulator will show an error. Make sure you have a ground.

After you have drawn the schematic, go to Simulate-> Run->Edit Simulation Command as in Fig 2.4.

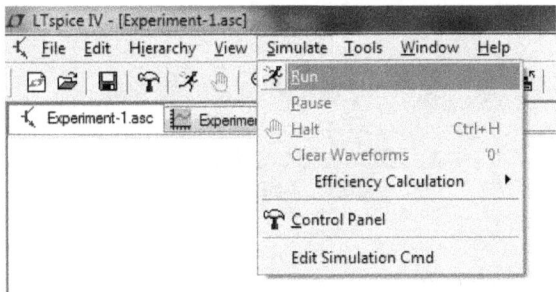

Fig 2.4: How to run option-1

The other way is to click on the run symbol (symbol- a running man) as shown in Fig 2.5.

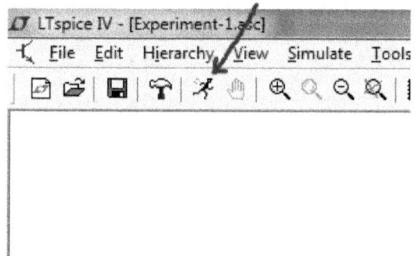

Fig 2.5: run symbol

In the Edit Simulation Command –>DC Sweep, The 1st source to be varied has been named as V2 as given in the schematic. The type of sweep chosen is linear. Other options are Octave, decade and list.

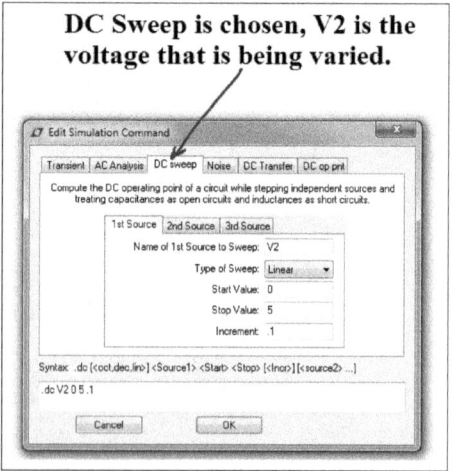

Fig 2.6: Selecting DC Sweep

Your First Circuit Simulation

When you click ok. A waveform window appears with no waveform in it.When you take your cursor near the resistor you can see two kinds of symbols as shown below.

 = Voltage measurement

 = Current measurement

When the current measurement option appears click on the resistor and the Waveform named I(R1) appears as shown in Fig 2.7.

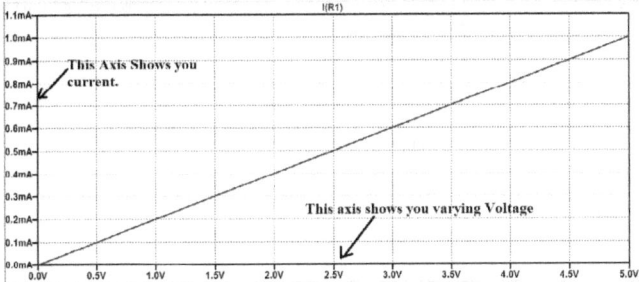

Fig 2.7: Voltage Vs Current plot for the circuit in Fig 2.3

From the above plot we can see that Voltage is directly proportional to the current through the resistor. The slope of the line in the graph is used to find the value of resistance.

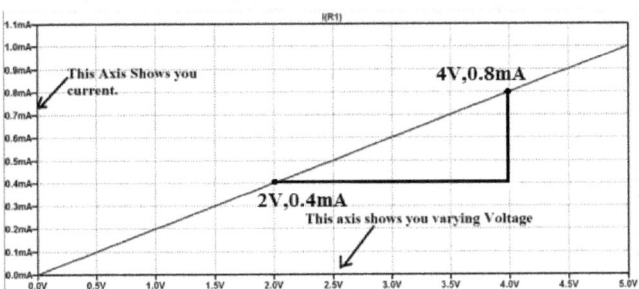

Fig 2.8: Calculation of Resistance from Voltage Vs Current plot

Calculation of Resistance from the Graph

$$\text{slope} = \frac{(0.8\text{mA} - 0.4\text{mA})}{(4\text{V} - 2\text{V})} = \frac{0.4\text{mA}}{2\text{V}} = 0.2\text{m}\frac{A}{V}$$

$$\text{Resistance from the graph} = \frac{1}{\text{slope}} = \frac{1}{0.2\text{mA}/V} = 5\text{k}\Omega$$

This is the value of resistance that we took in our Schematic. Thus Ohms law is verified from the graph.

****

Chapter 3   Diode

The concept of a Diode

Definition: A Diode is a two terminal semiconductor device that allows current to pass in one direction and blocks the current in the reverse direction. The two terminals are called Anode and Cathode. Anode is the terminal that is connected to P region of the diode and Cathode is the terminal that is connected to N region of the diode.

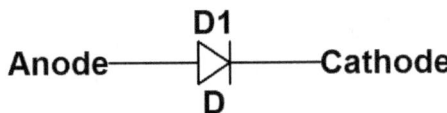

Fig 3.1: A Simple Diode

A Diode only conducts when the positive terminal of a supply voltage is connected to the Anode terminal of the Diode and negative of the supply is connected to the cathode terminal of the Diode. This is called forward biasing a Diode. In this case the diode works like a closed switch.

When the Anode is connected to the negative terminal of a battery and the cathode is connected to the positive terminal of the battery the diode does not conduct. This is called reversed biasing a Diode. In the reverse biased case the diode works like an open switch. Even if no voltage is applied obviously the diode acts like an open switch.

So the Diode is initially like a closed switch that requires a positive voltage to be applied to the anode with respect to its cathode in order to make it a closed switch. That is, we apply forward biasing to close the switch.

We can visualize a Diode like an open bridge that needs a force to close. Once we have applied a certain amount of force on the bridge. The bridge gets closed and current can now cross over to the other side of the bridge.

V-I Characteristics of a Diode

Voltage Vs Current characteristics of a Diode gives insights into the behavior of a Diode when a biasing voltage is applied. For forward biasing characteristics, connect the circuit components as shown in Fig 3.2 and choose DC sweep from the Edit Simulation Command.

Electronics Circuit SPICE Simulations with LTspice

Forward Bias

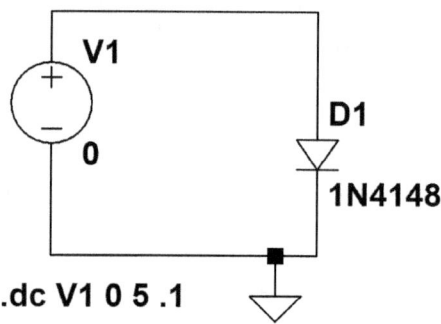

Fig 3.2: Forward biasing a Diode

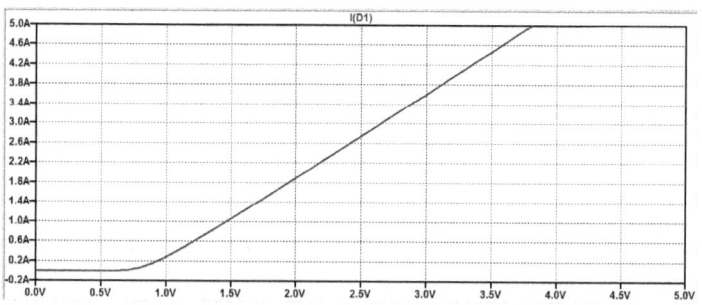

Fig 3.3: V-I Characteristics of a Diode in forward biasing

From Fig 3.3 we can observe that the current is almost negligible up to a forward bias voltage of 0.7 Volts. Post 0.7V the currents begins to rise sharply. This means that we need to apply more than 0.7 volts for the diode to work like a closed switch and conduct.

Diode

Reverse Bias

To get the reverse bias characteristics simply reverse the diode in the Fig 3.2 as shown in Fig 3.4(a). Fig 3.4(a) and Fig 3.4(b) essentially mean the same.

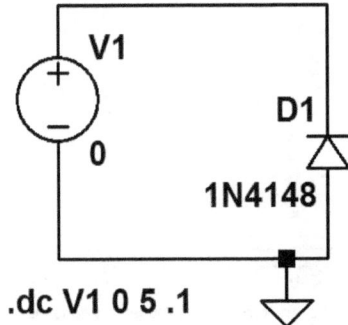

Fig 3.4(a): Reverse biasing a Diode (method-1)

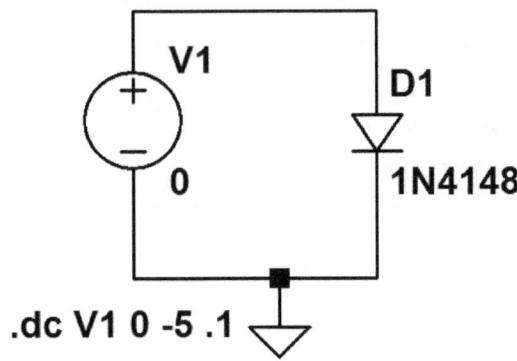

Fig 3.4(b): Reverse biasing a Diode (method-2)

Electronics Circuit SPICE Simulations with LTspice

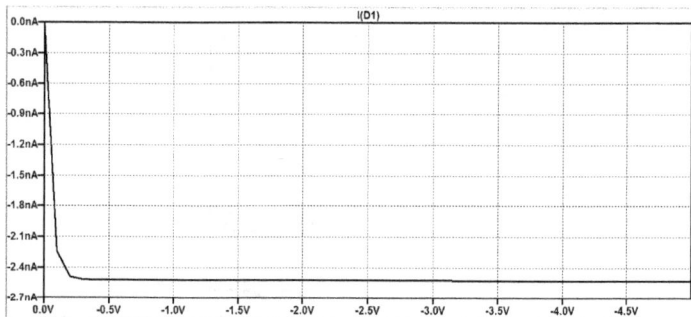

Fig 3.5: V-I Characteristics of a Diode in reverse biasing

From Fig 3.5 we can observe that a negligible amount of current in the range of nanoAmperes flows in the Diode because of the minority carriers. As we keep increasing the voltage the current rises initially for small values of voltages and then there is no further rise in the current as the voltage is increased further. This means that almost no current is flowing through the Diode and the Diode works like a open switch, i.e. it doesn't conduct.

Diode- Example Circuit

Circuit when the Diode is Forward Biased

Fig 3.6 shows a Circuit where the Diode is Forward biased. We apply KVL to the circuit.

Applying KVL to the circuit in Fig 3.6.

$10 - (10k)I - 0.7 = 0$

$$\Rightarrow I = \frac{10 - 0.7}{10k} \ A$$

$\Rightarrow I = 9.3 \times 10^{-4} \ A$

$\Rightarrow I = 0.93mA$

Diode

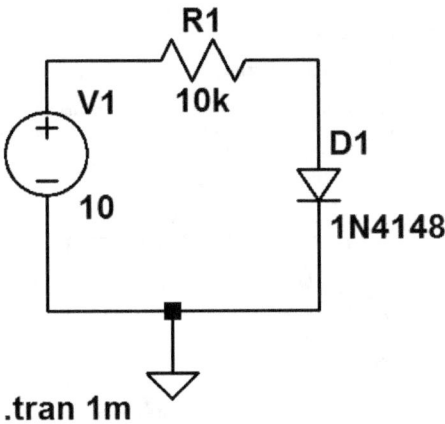

Fig 3.6: Circuit when the diode is Forward biased

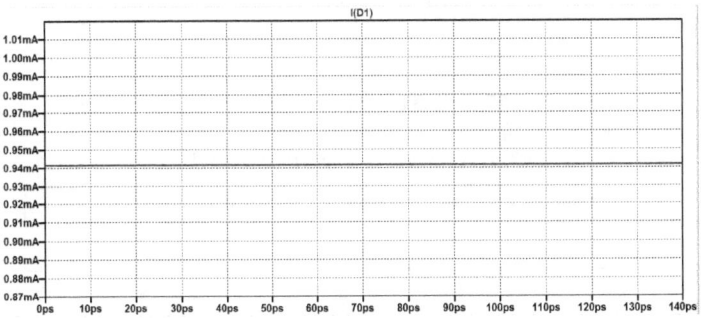

Fig 3.7: Transient response of the circuit in Fig 3.6

Fig 3.6 shows the plot of Transient Analysis. It can be seen that the diode current is constant at a value of 0.93mA.

Electronics Circuit SPICE Simulations with LTspice

Circuit when the Diode is Reverse Biased

When the diode is reverse biased, it acts like an open circuit and therefore a very negligible amount of current flows in the reverse direction to that of the forward biased case. This can be seen from the plot of Transient analysis as in Fig 3.9.

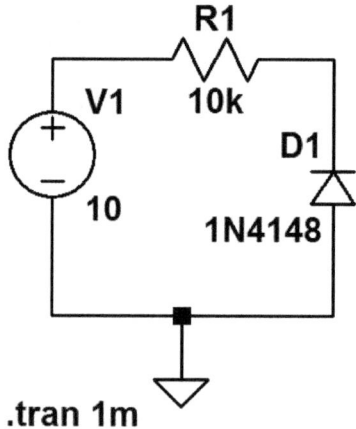

Fig 3.8: Circuit when the diode is reverse biased

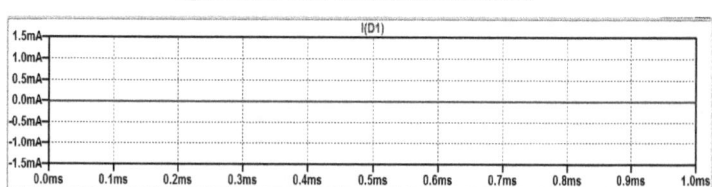

Fig 3.9: Transient response of the circuit in Fig 3.8

****

Chapter-4   Logic Gates Using Diodes

About the chapter
This chapter shows you simulations of 'OR' and 'AND' logic gates.Note that only these two logics can be implemented using Diodes and a resistor.

Contents of the chapter
1) OR gate using Diodes
2) AND gate using Diodes

We can implement simple logic circuits like 'AND' and 'OR' Gates using a combination of Diodes and a resistor.
Note: we will assume the diodes to be ideal in all our study throughout this chapter. This means that the diode will act like an ideal switch.

1) OR Gate

V1	V2	Vout
0	0	0
0	1	1
1	0	1
1	1	1

Table 4.1: Truth Table - OR gate

We have two inputs V1 and V2 and the output is Vout.The truth table of OR gate can be seen from Table 4.1. A two input OR gate can be implemented using a combination of two diodes and a resistor as shown in Fig 4.1.

Case 1: V1= 0V, V2=0V
If V1 is low(0), the diode D1 is reverse biased and no current flows through the resistor R1 because of V1. If V2 is low(0) then in that case again the diode D2 also reverse biased and no current flows through the resistor R1 because of V2. Since there is no current flowing through the resistor R1, there is no drop across itandthus Vout remains at low(0).

Case 2: V1= 0V, V2 = 5V
Since V1 is low(0)the diode D1 Is reverse biased and no current flows through the resistor R1 because of V1. As (V2) is high(5V), the diode D2 will become forward biased. As a result it will conduct and current will flow through the resistor R1 and a drop of 5V will occurs across the resistor R1. As a result we can measure a high(5V) at Vout.

Electronics Circuit SPICE Simulations with LTspice

Case 3: V1= 5V, V2 = 0V
Since V2 is low(0) the diode D2 IS reverse biased and no current flows through the resistor R1 because of V2. As V1 is high(5V), the diode D1 is forward biased. As a result it will conduct and current will flow through the resistor R1 and a drop of 5V will occur across the resistor R1. As a result we can measure a high(5V) at Vout.

Case 4: V1= 5V, V2 = 5V
Since V1 is high(5)the diode D1 Is forward biased a current flows through the resistor R1 because of V1. As (V2) is also high(5V), the diode D2 will also become forward biased. As a result it will alsoconduct and a current will flow through the resistor R1 and a drop of 5V will occurs across the resistor R1.Since the two supply voltages V1 and V2 are connected in parallel the voltage drop across R1 can't Exceed 5V. As a result a high(5V) at Vout can be measured.

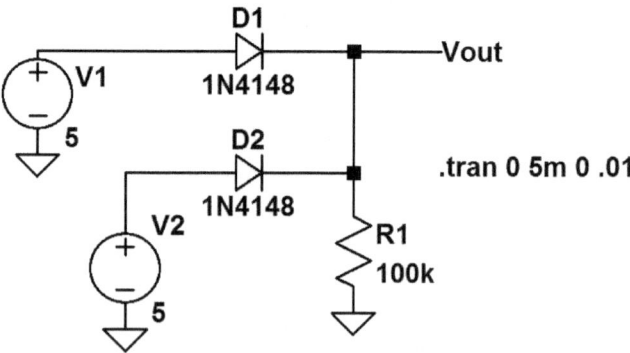

Fig 4.1: OR Logic-Circuit Diagram with both inputs high(5)

Logic Gates Using Diodes

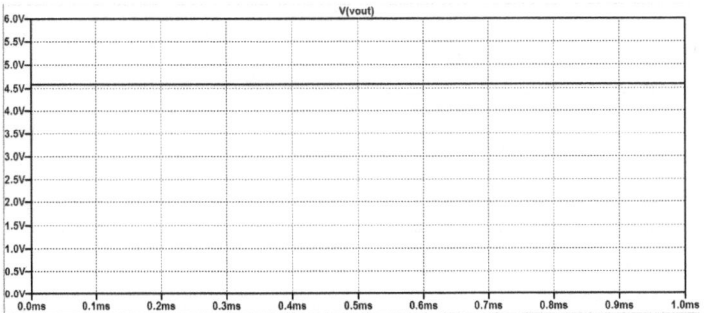

Fig 4.2: Transient Analyis: Or Logic circuit when both the Inputs are high(5)

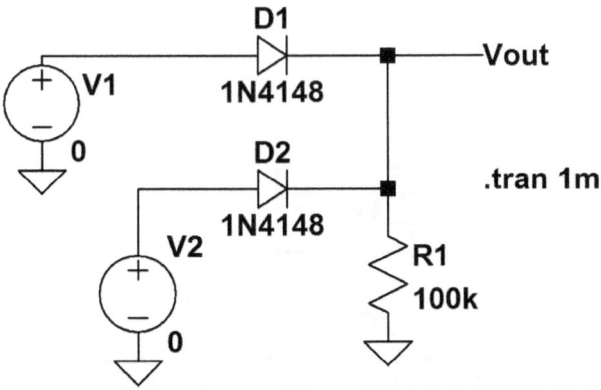

Fig 4.3: OR Logic-Circuit Diagram with both inputs low(0)

Electronics Circuit SPICE Simulations with LTspice

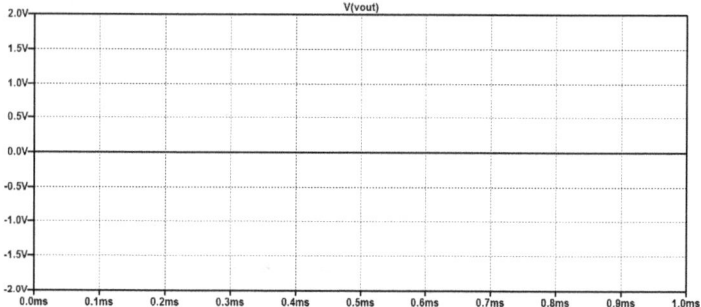

Fig 4.4: Transient Analysis: Or Logic circuit when both the inputs are low(0)

2) And Gate

V1	V2	Vout
0	0	0
0	1	0
1	0	0
1	1	1

Table 4.2: Truth Table - AND gate

We have two inputs V1 and V2 and the output is Vout. The truth table of AND gate can be seen from Table 4.2. A two input AND gate can be implemented using a combination of two diodes and a resistor as shown in Fig 4.5.

Case 1: V1= 0V, V2=0V
If V1 is low(0), the diode D1 is forward biased because of the V3. As a result there is a resistor free path available for the current to flows through the resistor R1. If V2 is low(0) then in that case again the diode D2 is also forward biased because of the V3and current flows through the resistor R1 .And thus a voltage drop of 5V occurs across the resistor R1 and Vout remains at low(0).

Case 2: V1= 0V, V2 = 5V
Since V2 is high(5) the diode D2 is reverse biased because of V3 and no current flows through the resistor R1 .But as V1 is low(0), the diode D1 is forward biased because of the V3. As a result there is a resistor free path available for the current to flows through the resistor R1.And thus a voltage drop of 5V occurs across the resistor R1 and Vout remains at low(0).

Case 3: V1= 5V, V2 = 0V
Since V1 is high(5) the diode D1 is reverse biased because of V3 and no current flows through the resistor R1 . But as V2 is low(0), the Diode D2is forward biased because of the V3. As a result there is a resistor free path available for the current to flows through the resistor R1. And thus a voltage drop of 5V occurs across the resistorR1 and Vout remains at low(0).

Case 4: V1= 5V, V2 = 5V
Since V2 is high(5) the diode D2 is reverse biased because of V3 and no current flows through the resistor R1. Again Since V2 isalso high(5) the diode D1 is reverse biased because of V3 and no current flows through the resistor R1.As both the inputs are set to a high voltage, both the diodes are reverse biased. As a result there will be no flow of current through the resistor R1. Since no current is flowing through the resistor, there will be no drop across it and a full 5V (present at V3) appears at Vout.

Summary- AND Gate

Lets say if one of the inputs(V1) is set to low(0)then that particular diode (D1) becomes forward biased because of the V3 voltage. As a result there is a resistor free path available for the current to flow through this diode. And thus a voltage drop of 5V occurs across the as current flowsacross resistor R1. The other Diode(D2) is reverse biased(like an open circuit) and as a result does not conduct any current. So if any of the inputs is set to low(0), the output is low(0) and this defines And gate as per the table 4.2.
If both the inputs are set to a high voltage, then in this case both the diodes are reverse biased. As a result there will be no flow of current through the resistor R1 and since no current is flowing through the resistor, there will be no drop across it and a full 5V (present at V3) will appear at Vout.

Electronics Circuit SPICE Simulations with LTspice

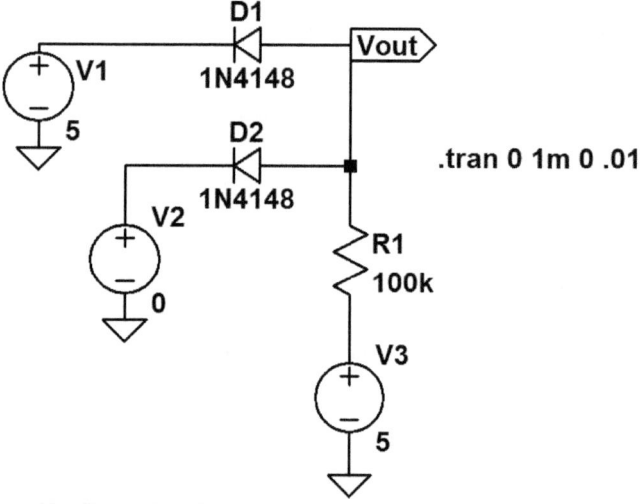

Fig 4.5: AND Logic-Circuit Diagram when one of the inputs is low(0)

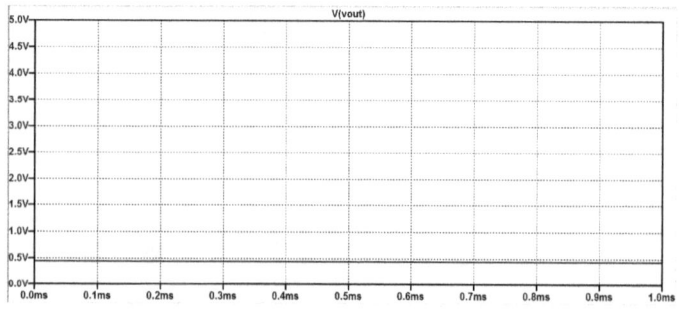

Fig 4.6: Transient Analysis: AND Logic circuit when one the Inputs is low(0)

Logic Gates Using Diodes

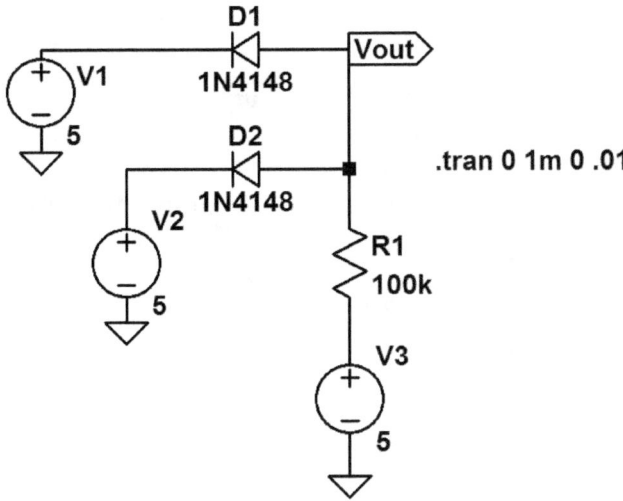

Fig 4.7: AND Logic-Circuit Diagram when both inputs are high(5)

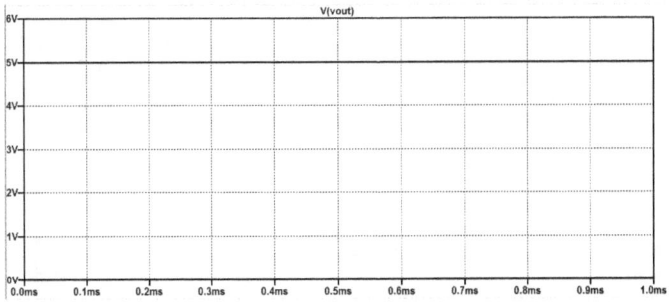

Fig 4.8: Transient Analysis: And Logic circuit with both the Inputs are high(5)

Electronics Circuit SPICE Simulations with LTspice

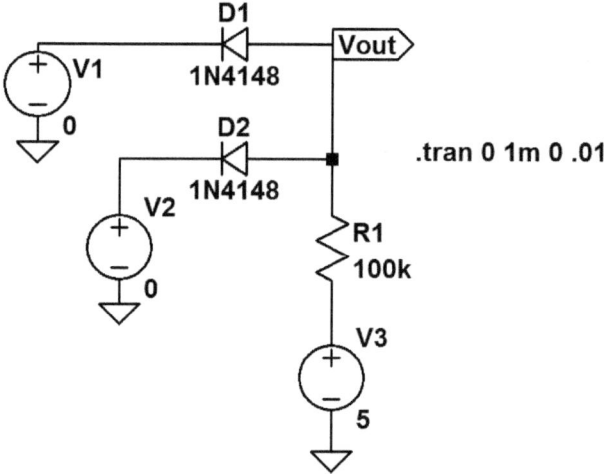

.tran 0 1m 0 .01

Fig 4.9: AND Logic-Circuit Diagram when both inputs are low(0)

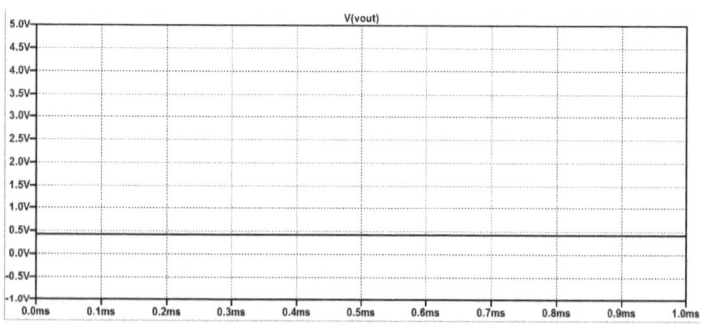

Fig 4.10: Transient Analysis: And Logic circuit when both the inputs are low(0)

****

Chapter 5   Clipper Circuit Using Diode

About this chapter

This chapter shows you various simulations of clipper circuits using the Diode.

Contents of the chapter:

1) Series diode clipping circuits

2) Parallel diode clipping circuits

Introduction

A clipper circuit removes a particular portion of the input signal and gives an output signal which is clipped. Clipping a signal is similar to cutting a cloth with a scissor. The undesired signal is simply clipped off. Diode is widely used in the design of various clipper circuits. There are two types of diode clipper circuits.The series diode clipper circuit wherein the diode is connected in between the input and the output and the parallel diode clipper circuit wherein the diode is connected in parallel with the output. The next section discusses the operation of both these clipper circuits in detail.

1) Series diode clipper circuit

There are many circuit configurations of series diode clipper circuits. Here in this chapter we discuss two basic types of series diode clipper circuit, the negative half clipping circuit and the positive half clipping circuit.

Circuit diagram - negative half clipping circuit

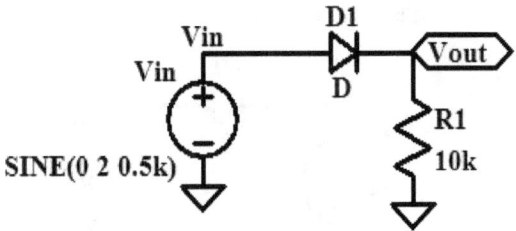

Fig 5.1: Circuit diagram- negative half clipping circuit

Electronics Circuit SPICE Simulations with LTspice

Simulation results- negative half clipping circuit

Command for analysis: .tran 0 20m 0 0.01m

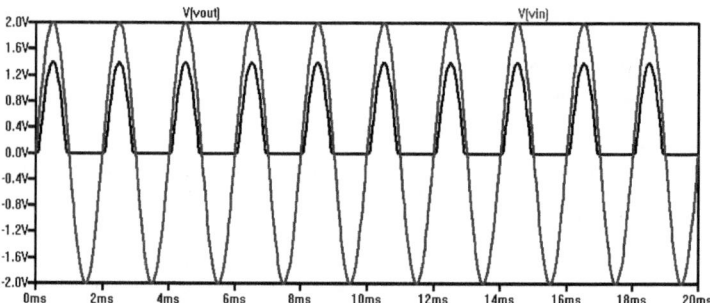

Fig 5.2: Simulation result of negative half clipping circuit

Analysis - negative half clipping circuit

During the positive half cycle of the sinusoidal input signal the diode D1 turns ON after 0.7 V and passes the input signal to the output. There is a voltage drop of 0.7 V across the diode so the output level is below 0.7 V with respect to input signal. During negative half cycle of input signal the diode is reverse biased, so at output the signal is 0 V. From Fig 5.3, it can be seen that the negative half of the input sinusoidal signal is clipped.

Circuit diagram - positive half clipping circuit

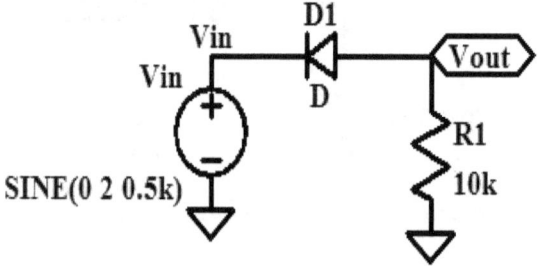

Fig 5.4: Circuit diagram- positive half clipping circuit

Clipper Circuit Using Diode

Simulation results - positive half clipping circuit

Command for analysis: .tran 0 20m 0 0.01m

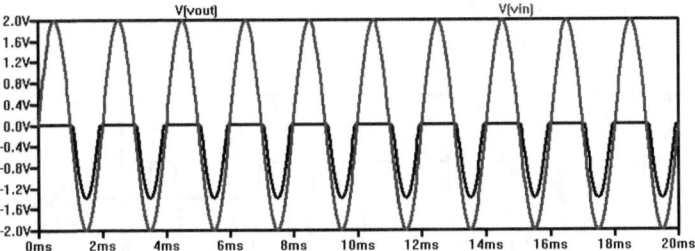

Fig 5.5: Simulation results - positive half clipping circuit

Analysis- positive half clipping circuit

During the negative half cycle of the input signal the diode turns ON after 0.7 V and passes the input signal to the output. There is 0.7 V voltage drop across diode so the output level is 0.7 V lesser with respect to the input signal. During positive half cycle of input signal the diode is reverse biased, so at output the signal is 0 V.

2) Parallel diode clipping circuit

In this configuration as well many circuits are possible but here we will discuss about two types of circuits, the positive half clipping circuit and the negative half clipping circuit.

Circuit diagram- positive half clipping circuit

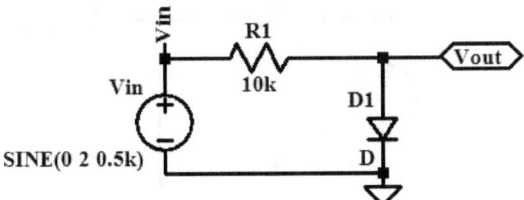

Fig 5.6: circuit diagram - positive half clipping circuit

Electronics Circuit SPICE Simulations with LTspice

Simulation results- positive half clipping circuit

Command for analysis: .tran 0 20m 0 0.01m

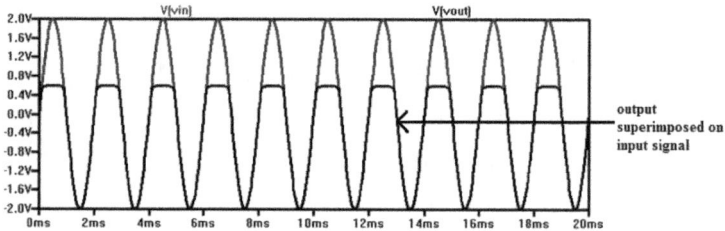

Fig 5.7: Simulation results - positive half clipping circuit

Analysis - positive half clipping circuit

During positive half cycle of input signal the diode is forward biased so the output voltage is equal to the voltage drop across diode i.e. 0.7 V approximately. During negative half cycle the diode is reverse biased so the input signal passes through the resistor to the output, so the output is equal to the input signal.

Circuit diagram - negative half clipping circuit

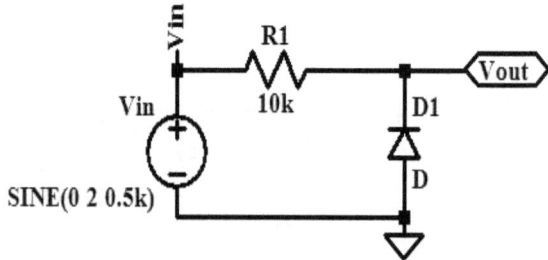

Fig 5.8: Circuit diagram - negative half clipping circuit

Clipper Circuit Using Diode

Simulation results - negative half clipping circuit

Command for analysis: .tran 0 20m 0 0.01m

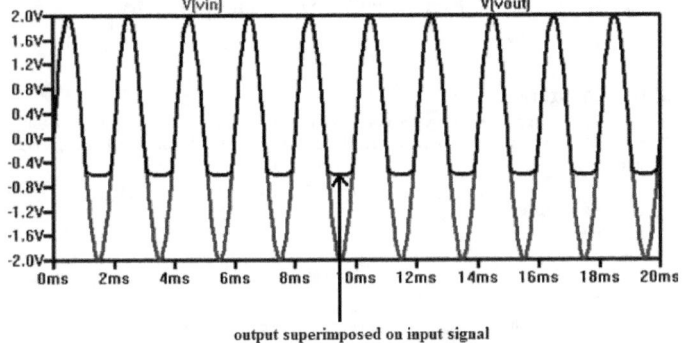

output superimposed on input signal

Fig 5.9: Simulation results - negative half clipping circuit

Analysis - negative half clipping circuit

During the negative half cycle of input signal the diode is forward biased so the output voltage is equal to the voltage drop across the diode i.e. 0.7 V approx. During positive half cycle the diode is reverse biased so the input signal passes through the resistor to the output, so the output is equal to the input signal.

Thus in a series clipper circuit when the diode is reverse biased it clips that cycle of the input signal and in parallel clipper circuit when diode is reverse biased it passes the input signal to the output.

****

Chapter 6　　　　Clamper Circuit Using Diode

About this chapter
In this chapter we discuss about the design and simulation of clamper circuit using diode. Again we will discuss about types of clamper circuits, the positive clamper circuit and the negative clamper circuit.

Contents of the chapter
1) Positive clamper
2) Negative clamper

Introduction
A clamper is a circuit that shifts the entire signal vertically up or down such that the DC Value of the signal changes. The entire signal is shifted to a new reference value and the signal is neither clipped nor limited.   A typical diode clamper circuit consists of a diode and a capacitor. The capacitor and a diode are connected in parallel with the load. Fig 6.1 shows the circuit diagram of a positive clamper circuit. Fig 6.2 shows the transient analysis of a positive clamper circuit. It can be seen that the entire signal has shifted upwards vertically. Similarly Fig 6.3 shows the simulation results of a negative clamper circuit. It can be seen that the entire signal has shifted downwards vertically.

1) Positive clamper
Circuit diagram

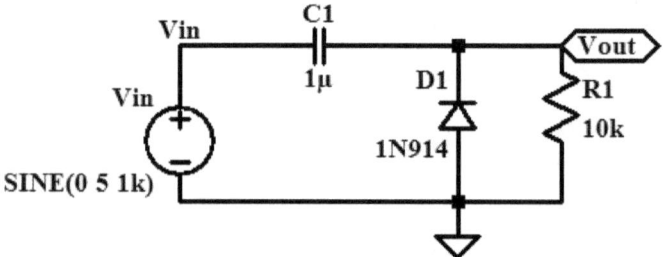

Fig 6.4: Circuit diagram of positive clamper circuit

Clamper Circuit Using Diode

Working

During the positive half cycle of the sinusoidal input signal the diode D1 is OFF due to reverse bias and thus the capacitor C1 charges through the register R1. The capacitor charges up to the supply voltage i.e. 4.3 V and holds the charge. During this cycle the output is equal to the input signal. During the negative cycle of the input signal the diode D1 is ON, so the output voltage is equal to the sum of input voltage and voltage across the capacitor. Note that this working is for one cycle of the input signal.

After one complete cycle of the input signal, the output signal is clamped approx. to 4.3 V with respect to ground level because after one complete cycle of the input signal, during positive and negative cycle both, the output voltage is equal to the sum of input signal and voltage across the capacitor.

The output voltage is clamped to 4.3 V approximately. This is because the diode has a voltage drop of 0.7 V. As a result the maximum voltage across the capacitor during charging phase is only 4.3 V instead of 5 V.

Simulation Result

Command for analysis: .tran 0 20m 0 0.01m

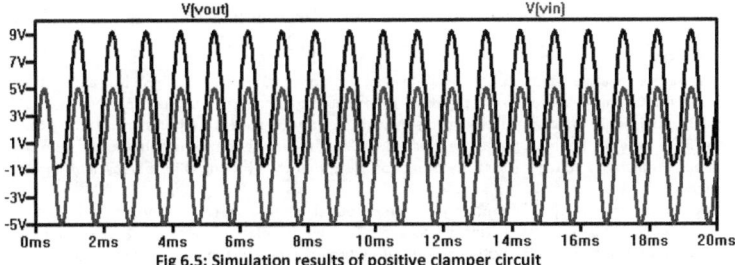

Fig 6.5: Simulation results of positive clamper circuit

Electronics Circuit SPICE Simulations with LTspice

2) Negative clamper
Circuit diagram

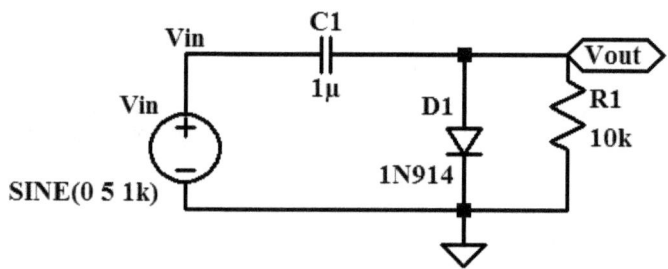

Fig 6.6: Circuit diagram of negative clamper circuit

Working

During the negative half cycle of the input signal the diode is OFF due to reverse bias and so the capacitor C1 charges through the register R1. The capacitor charges up to the supply voltage i.e. -4.3 V and holds the charge. During this cycle the output is equal to input signal. During the positive cycle of the input signal the diode is ON, so the output voltage is equal to the sum of input voltage and voltage across the capacitor. Note that this working is for one cycle of input signal only.

After one complete cycle of input signal, the output signal is clamped to approximately -4.3 V with respect to ground level. This is because after one complete cycle of input signal, during positive and negative cycle both, the output voltage is equal to the sum of input signal and voltage across the capacitor.

Clamper Circuit Using Diode

Simulation Result
Command for analysis: .tran 0 20m 0 0.01m

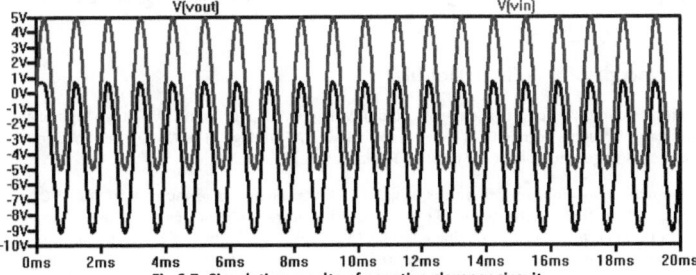

Fig 6.7: Simulation results of negative clamper circuit

****

Chapter 7   Voltage Doubler Circuit using Diodes

The concept of a Voltage Doubler

Definition: A Voltage Doubler is a specific kind of Voltage Multiplier circuit. A Voltage Multiplier circuit can produce a higher DC Voltage from a lower AC Voltage. The Output DC Voltage is typically an integer multiple of the peak value of AC Voltage.

For example an AC Voltage with a peak value of 20 Volts can be converted to a 40 Volts DC or 80 Volts DC. In this chapter we study the simplest Voltage multiplier circuit, the half wave Voltage Doubler. In a Half wave Voltage Doubler the Output DC Voltage is twice the maximum peak value of the AC Input. Fig 7.1 shows the circuit diagram of a typical Half-wave Voltage Double Circuit without the load.

Principle of operation

In the positive half cycle the of the Sinusoidal input, diode D2 acts as an open switch and the diode D1 acts as a closed switch. As a result the Capacitor C2 is charged to V1(max) and the capacitor C1 is charged to 2 V1(max). For example considering Fig 7.1, V1(max) = 10 V. Thus in the positive half cycle C2 is charged to 10V and C1 is charged to 20V.

In the negative half cycle the diode D2 acts as a closed switch and the diode D1 acts as a open switch. As a result the Capacitor C2 remains charged to V1(max) and the capacitor C2 starts discharging. For example considering Fig 7.1, in the negative half cycle C2 remains charged to 10V and C1 starts discharging. Since there is no load in the circuit, and the only path available for C1 to discharge is through C2 as a result the discharging is slow.

By the time C1 has barely discharged, the next positive cycle of the sinusoidal input arrives and C1 is again charged to 20V. Fig 7.2 shows the transient Analysis of Half-wave Voltage Doubler Circuit without the Load.

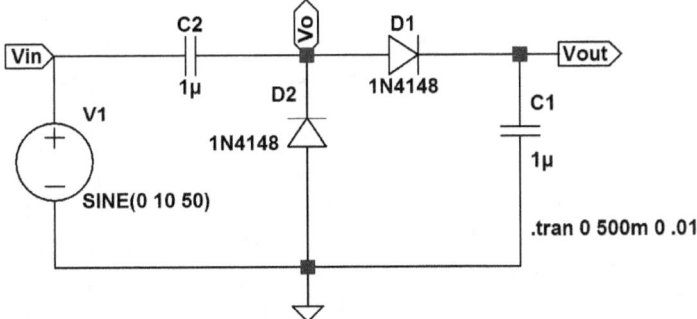

Voltage Doubler Circuit using Diodes

Fig 7.1: A typical Half-wave Voltage Doubler Circuit without the load

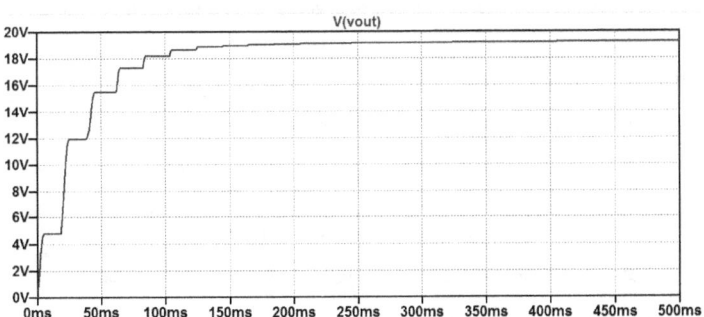

Fig 7.2: Transient Analysis: Half-wave Voltage Doubler Circuit without the Load

Electronics Circuit SPICE Simulations with LTspice

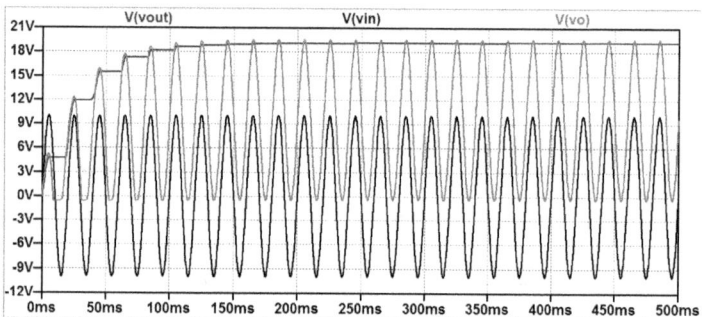

Fig 7.3: Transient Analysis: Voltage levels- Half-wave Voltage Doubler Circuit without the Load

Half-wave Voltage Doubler Circuit with the Load

Fig 7.4 shows the circuit diagram of a Half-wave Voltage Double Circuit with the load. Since there is a load now in the circuit, there is path available for C1 to discharge is through R1 as a result the discharging is faster. Fig 7.5 shows the transient Analysis of Half-wave Voltage Doubler Circuit with the Load. It can be seen that the output DC has a lower value now.

Fig 7.6 shows A Half-wave Voltage Doubler Circuit with the load = R1=10k. Fig 7.7 shows the Transient Analysis of the Half-wave Voltage Doubler Circuit with the load (R1=10k). It can be seen that the output DC has a even lower value now because of the lower R1 Value. As the value of R1C1 decides the performance of the circuit, these values must be chosen properly.

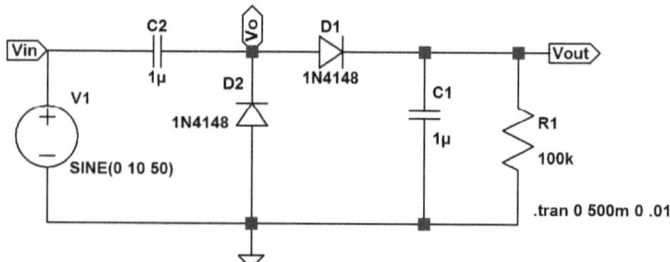

Fig 7.4: A Half-wave Voltage Doubler Circuit with the load

Voltage Doubler Circuit using Diodes

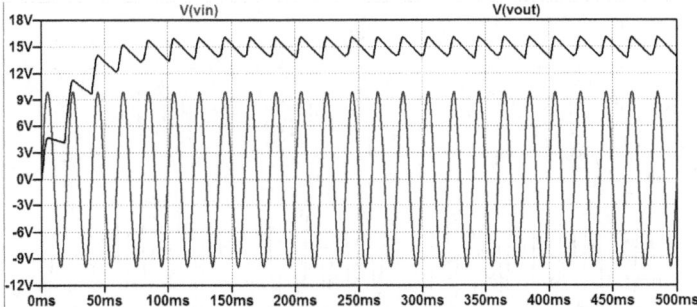

Fig 7.5: Transient Analysis: Half-wave Voltage Doubler Circuit with the load(R1=100k)

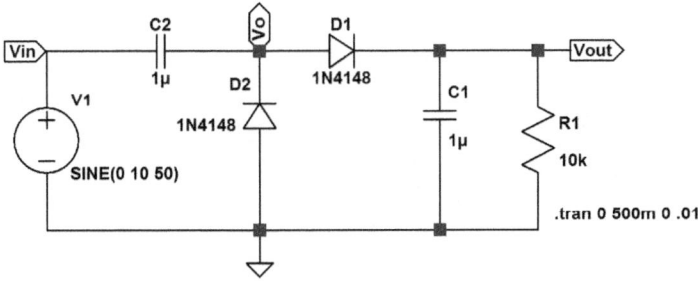

Fig 7.6: A Half-wave Voltage Doubler Circuit with the load(R1=10k)

Electronics Circuit SPICE Simulations with LTspice

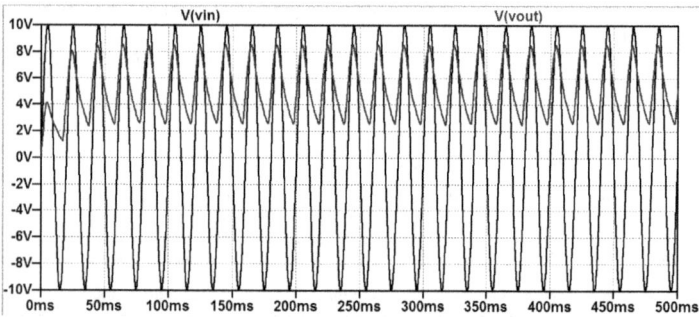

Fig 7.7: Transient Analysis: Half-wave Voltage Doubler Circuit with the Load(R1=10k)

****

Chapter 8   Bipolar Junction Transistor (BJT)

The concept of a BJT

BJT stands for Bipolar Junction Transistor (B= Bipolar, J= Junction, T= Transistor). The first terms says Bipolar (Bi = two, polar = having poles) which means that there are two kinds of polarities, i.e. charge carriers in a BJT namely holes and electrons. The second terms is Junction, BJT has two junctions, Emitter-Base junction and the Base-Collector Junction. A BJT can be either a NPN or a PNP. NPN means that a P region is sandwiched between two N regions. Similarly PNP means that a N region is sandwiched between two P regions. BJT has three terminals namely Emitter, Base and Collector as shown in figure 8.1. As the names suggest, Emitter emits electrons or holes, Base controls the flow of the electrons or holes and finally a fraction of electrons or holes are collected at the collector.

Definition:

BJT is a three terminal semiconductor device that can work like either an amplifier or a switch based on the region of its operation.

BJT-Working principle and Simulation

Since for both the NPN and PNP transistors the concepts are the same, we will focus only on one of these, NPN. All the simulations in this book will be focused on NPN transistors only.

When a NPN transistor is in the unbiased state it can be seen like two diodes connected back to back as shown below in Fig 8.1. But this is not an accurate model of a BJT and must only be used to understand BJT in a crude manner.

A BJT starts working when suitable biasing is applied at its terminal. In order to make the transistor work the Base-emitter diode must be first forward biased (Base terminal placed at a higher potential than the emitter terminal). This causes a current to flow through the base emitter junction as the Base-emitter junction is now forward biased. But again there is no flow of current across the base-collector junction. A high voltage needs to be applied to the collector terminal so that current starts flowing across the Base-Collector junction. This makes the Base collector junction reverse biased. Only a small amount of electrons combine with the holes present in the base region, most of the electrons thus reach the collector terminal. A large number of electrons from the emitter reach the collector primarily because the base region is lightly doped and is thinner than both the emitter and collector. A BJT is designed in such a manner that the flow of current can be controlled using the base i.e. the base voltage can be adjusted such that a required amount of current can flow through the BJT.

Electronics Circuit SPICE Simulations with LTspice

Accordingly a BJT can work in four different modes or regions.
1) Cut off: The BJT acts like an open switch.
2) Active: There is a current flow from the emitter to the collector that is proportional to the current flowing into the base terminal, i.e. $I_c = \beta I_b$ where β is a constant for a particular BJT and is called the DC Current gain.
3) Saturation: The BJT acts like a closed switch.
4) Reverse Active: Practically not very useful.

A BJT can work in three different configurations namely
1) Common Emitter
2) Common Base
3) Common Collector

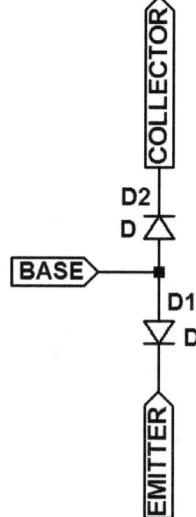

Fig 8.1: A NPN Transistor as two Diodes

Bipolar Junction Transistor

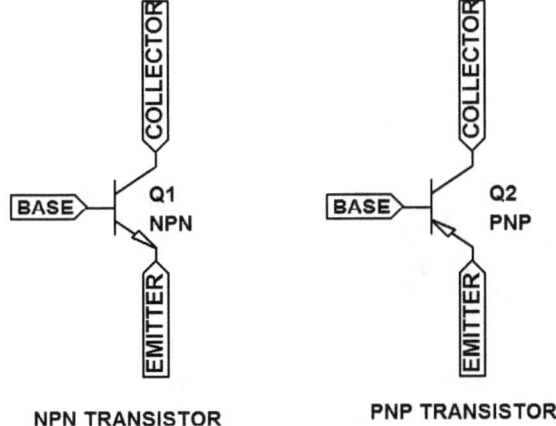

NPN TRANSISTOR   **PNP TRANSISTOR**

Fig: 8.2 PNP and NPN type BJTs - Schematic symbols

In this chapter we focus only on the simulation of common emitter configuration as it is the most commonly used configuration. In a CE configuration the emitter terminal of the BJT is grounded as shown in fig 8.3.

The behavior of a BJT in CE configuration can be fully explained with the help of two Characteristics, namely

1) Input characteristics - Also called base characteristics in which V_{BE} is varied and I_B is measured for different values of V_{CE}.

2) Output Characteristics - Also called collector characteristics in which V_{CE} is varied and I_C is measured for different values of I_B.

43

Electronics Circuit SPICE Simulations with LTspice

The Transistor Q1

Transistor 2N2222 has been chosen for this simulation. To choose a particular transistor, right click on the transistor as shown in Fig 8.3.

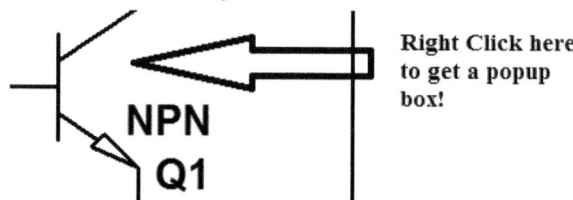

Fig 8.3: Selecting a particular NPN Transistor

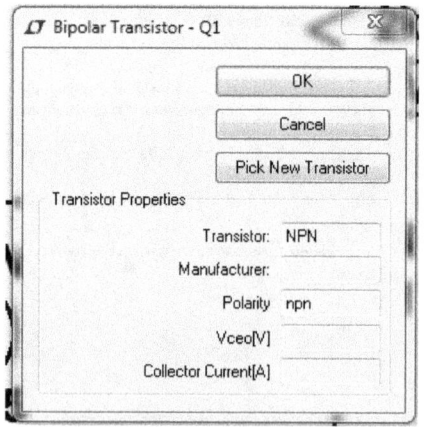

Fig 8.4: Pick New Transistor pop up

When the pop up window appears as in Fig 8.4, click on - pick new transistor. This will show you a set of transistors from where you can choose one. You can also look at the features of a particular transistor and its Spice Model file.

Bipolar Junction Transistor

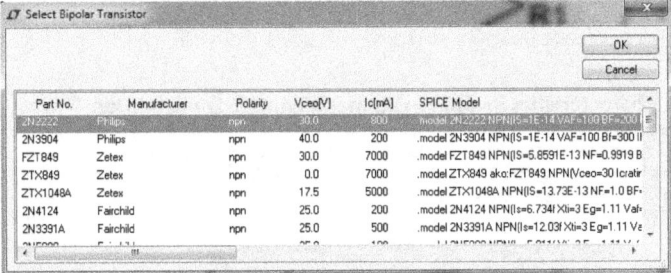

Part No.	Manufacturer	Polarity	Vceo[V]	Ic[mA]	SPICE Model
2N2222	Philips	npn	30.0	800	.model 2N2222 NPN(IS=1E-14 VAF=100 BF=200 I
2N3904	Philips	npn	40.0	200	.model 2N3904 NPN(IS=1E-14 VAF=100 Bf=300 II
FZT849	Zetex	npn	30.0	7000	.model FZT849 NPN(IS=5.8591E-13 NF=0.9919 B
ZTX849	Zetex	npn	0.0	7000	.model ZTX849 aka:FZT849 NPN(Vceo=30 Icratir
ZTX1048A	Zetex	npn	17.5	5000	.model ZTX1048A NPN(IS=13.73E-13 NF=1.0 BF=
2N4124	Fairchild	npn	25.0	200	.model 2N4124 NPN(Is=6.734I XIi=3 Eg=1.11 Vaf=
2N3391A	Fairchild	npn	25.0	500	.model 2N3391A NPN(IS=12.03I Xti=3 Eg=1.11 Va

Fig 8.5: Select Bipolar Transistor window.

Fig 8.6: Select Bipolar Transistor window- expanded view

From Fig 8.6, we can see the 2N2222-Spice Model File as:

2N2222- SPICE Model

.model 2N2222 NPN(IS=1E-14 VAF=100 BF=200 IKF=0.3 XTB=1.5 BR=3 CJC=8E-12 CJE=25E-12 TR=100E-9 TF=400E-12 ITF=1 VTF=2 XTF=3 RB=10 RC=.3 RE=.2 Vceo=30 Icrating=800m mfg=Philips)

Some of the important SPICE parameters are given below:

BF = Forward active current gain = β = 200

IS = Transport saturation current = 10^{-14}

VAF = Forward mode Early voltage = 100

BR = Reverse active current gain = 3

Input Characteristics of a BJT in Common Emitter Configuration

The input characteristics are just like a diode characteristics in the forward bias condition. This is simply because emitter base junction is just like a PN junction diode. For the input characteristics, connect the BJT as shown in Fig 8.7 and choose DC sweep from the Edit Simulation Command.

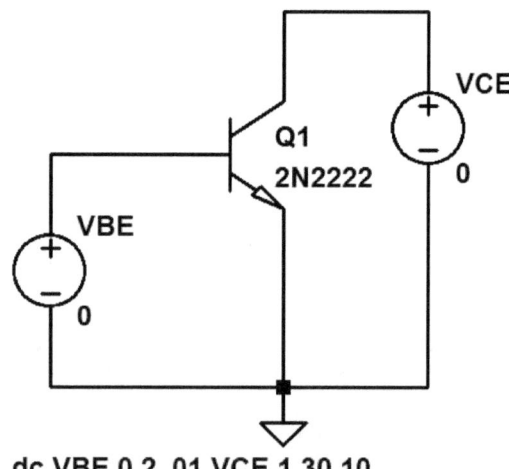

.dc VBE 0 2 .01 VCE 1 30 10

Fig 8.7: BJT in Common Emitter Configuration

Bipolar Junction Transistor

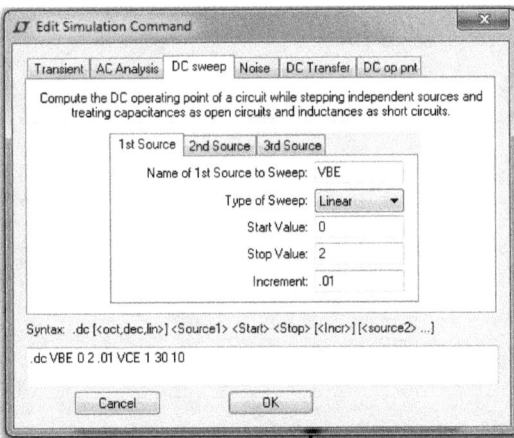

Fig 8.8: Setting the first Voltage source sweep values

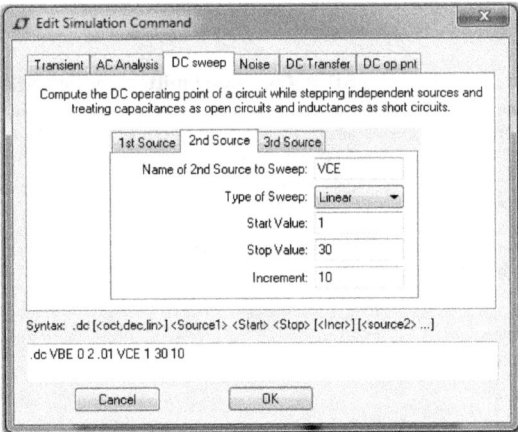

Fig 8.9: Setting the second Voltage source sweep values

Electronics Circuit SPICE Simulations with LTspice

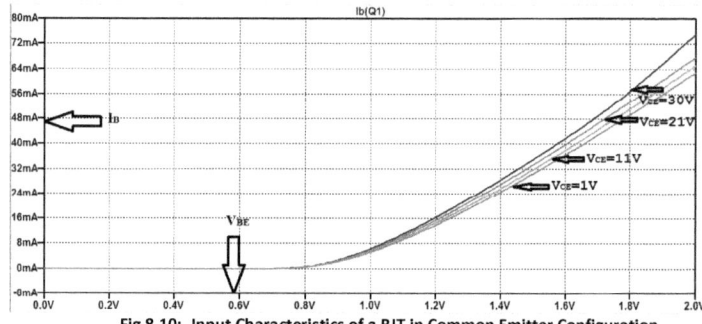

Fig 8.10: Input Characteristics of a BJT in Common Emitter Configuration

From Fig 8.10 we can observe that the current is almost negligible up to a forward bias voltage of 0.7 Volts. Post 0.7V the currents begins to rise sharply. This means that we need to apply more than 0.7 volts for the Base-emitter junction to work like a closed switch and conduct.

Output Characteristics of a BJT in Common Emitter Configuration

The output characteristics shows the nature of I_C when V_{CE} is varied keeping $_{I_B}$ as a constant value. The output characteristics are initially rising in nature and then become almost constant as V_{CE} is swept to higher values. For the Output characteristics, connect the BJT as shown in Fig 8.11 and choose DC sweep from the Edit Simulation Command. Note that the Voltage Source across the base and emitter terminals has now been replaced by a Current Source.

The output characteristics of a BJT in Common Emitter configuration has three regions of operation as shown in Fig 8.14:
1) Active Region
2) Cutoff Region
3) Saturation region

Bipolar Junction Transistor

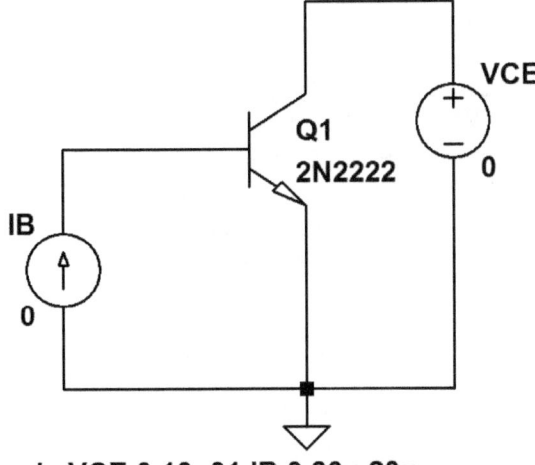

.dc VCE 0 10 .01 IB 0 80u 20u

Fig 8.11: BJT in Common Emitter Configuration (Circuit for output characteristics)

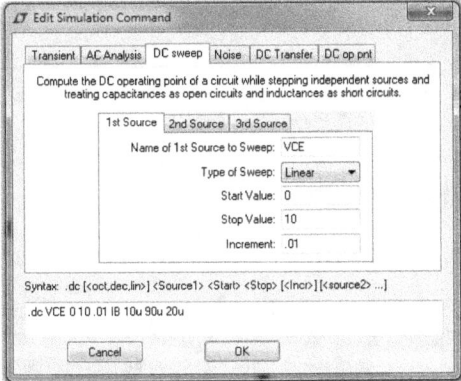

Fig 8.12: Setting the Voltage source sweep values

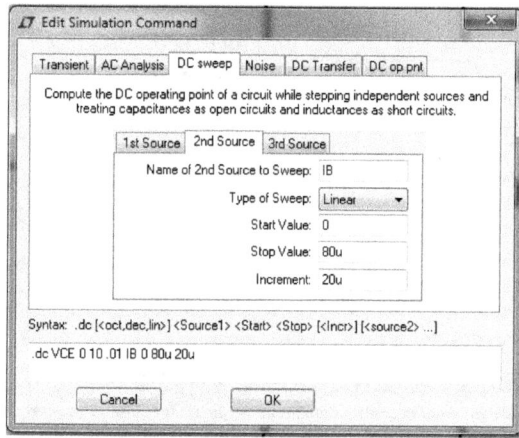

Fig 8.13: Setting the current source sweep values

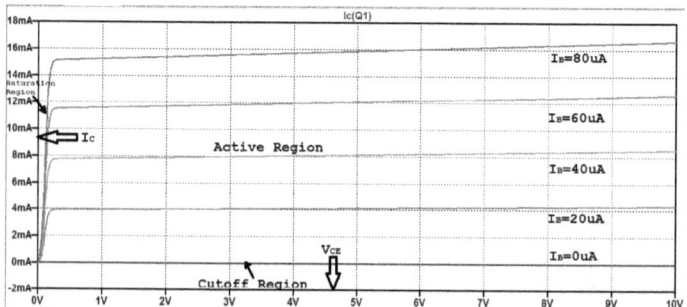

Fig 8.14: Output Characteristics of a BJT in Common Emitter Configuration

****

CHAPTER-9   NOT Gate using a NPN Transistor

About the chapter

This chapter shows you the simulations of a NOT Gate using a transistor and resistors. A NOT Gate can be designed using a NPN transistor and two resistors.

NOT Gate

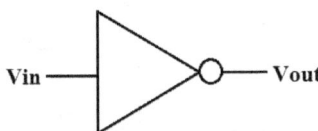

Fig 9.1: Symbol – NOT gate

Vin	Vout
0	1
1	0

Table 9.1: Truth Table - NOT gate

In a NOT gate, if the input is low (0V) then the output is high (5V) and vice versa. The truth table of NOT gate can be seen from Table 9.1. In Table 9.1 we have the input as V1 (Vin) and the output is Vout which is measured at the collector of the NPN transistor.

Operation of a NOT Gate using a NPN Transistor

Case 1: V1 = 5V

If V1 is high(5), the transistor Q1 is turned ON. As a result a current starts to flow between the collector and the emitter terminals of the transistor. This leads to a collector current(Ic) flowing through the resistor R1. Thus a drop of 5V occurs across the resistor R1 and the output voltage Vout becomes low(0). Fig 9.2 shows the schematic diagram of NOT Gate using NPN Transistor.

Electronics Circuit SPICE Simulations with LTspice

Fig 9.3 shows Transient Analysis of the NOT Gate where we can clearly see that the output is constant at a value of about 0V.

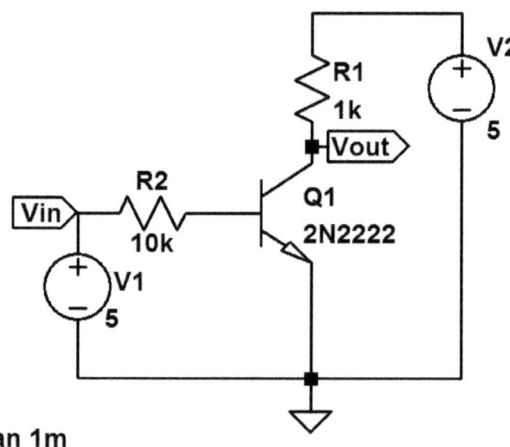

.tran 1m

Fig 9.2: Schematic Diagram – NOT Gate using NPN Transistor when the input is 5V

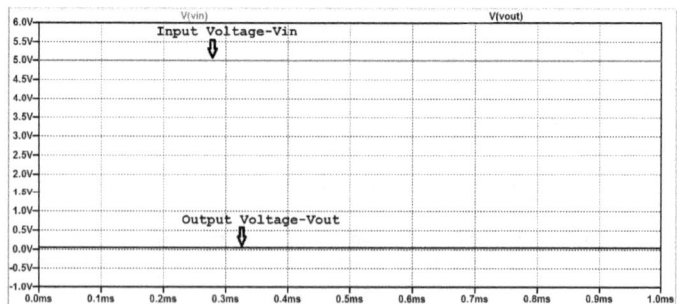

Fig 9.3: Transient Analysis - NOT Gate when the input is 5V

NOT Gate using a NPN Transistor

Case 2: V1= 0V

If V1 is low(0), the transistor Q1 is turned OFF. As a result no current flows between the collector and the emitter terminals of the transistor. As a result there is no voltage drop across the resistor R1. The entire voltage V2 appears at Vout and thus Vout becomes high(5). Fig 9.4 shows the schematic diagram of NOT Gate using NPN Transistor. Fig 9.5 shows Transient Analysis of the NOT Gate where we can clearly see that the output is constant at a value of about 5V.

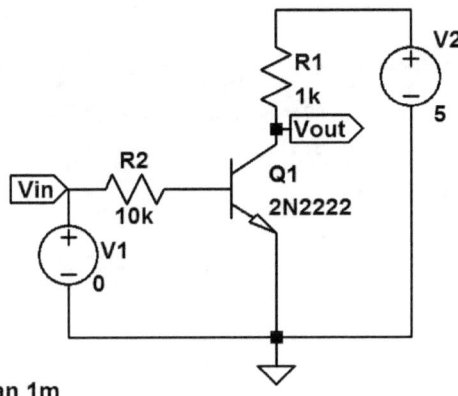

.tran 1m

Fig 9.4: Schematic Diagram – NOT Gate using NPN Transistor when the input is 0V

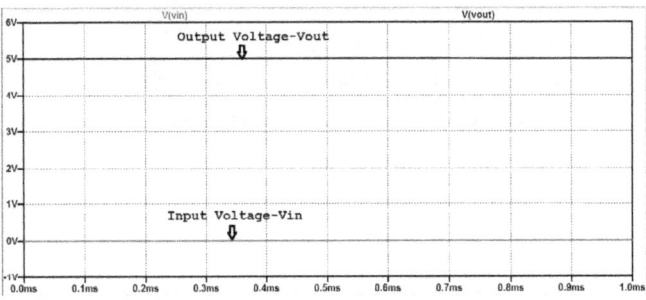

Fig 9.5: Transient Analysis - NOT Gate when the input is 5V

Electronics Circuit SPICE Simulations with LTspice

DC Sweep Analysis

For the DC Sweep Analysis, choose DC sweep from the Edit Simulation Command and set the Voltage source V1 as shown in Fig 9.6.From DC Sweep Analysis we can observe that when the input voltage is low, the output voltage is high and vice versa.

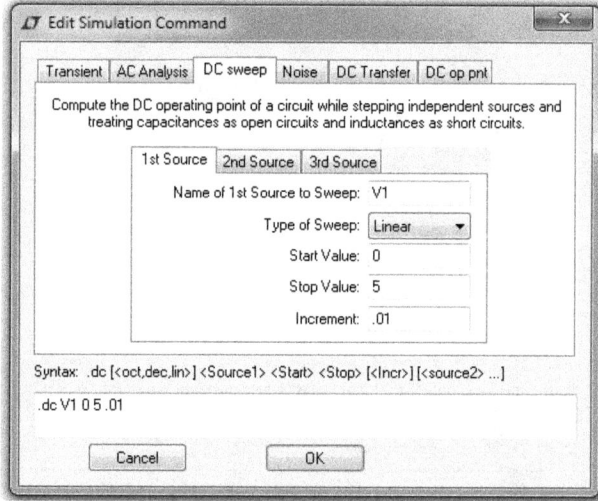

Fig 9.6: Setting the Input Voltage source sweep values

NOT Gate using a NPN Transistor

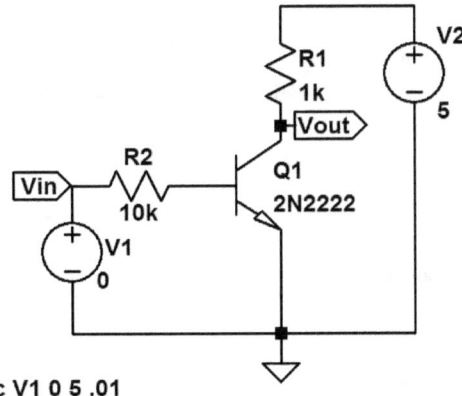

.dc V1 0 5 .01

Fig 9.7: DC Sweep Analysis- NOT Gate using NPN Transistor

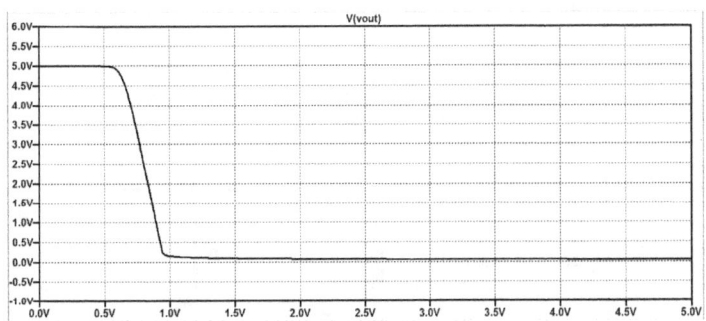

Fig 9.8: DC Sweep Analysis- Vout Vs Vin

****

Chapter 10   Common Emitter Amplifier

About the chapter

This chapter discusses about the simulation of common emitter amplifier with the aim of finding the DC operating point and gain of the amplifier.

Contents of the chapter

1) Circuit diagram

2) Principle of Operation

3) Operating point analysis

4) AC analysis

5) Transient analysis

1) Circuit Diagram

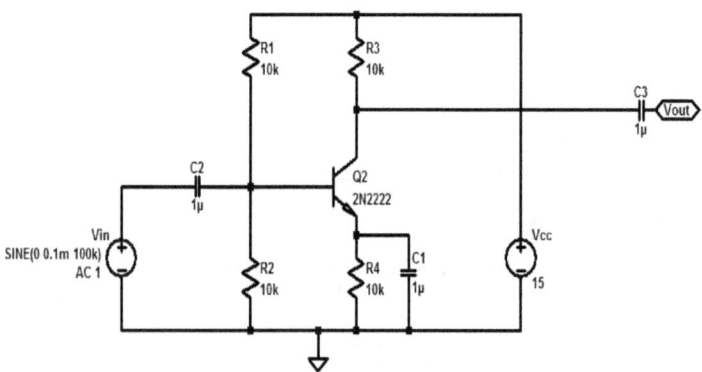

Fig 10.1: circuit diagram of a common emitter amplifier with typical values.

Common Emitter Amplifier

Figure 10.2 shows the circuit diagram of a common emitter amplifier with some typical values of resistors and capacitors. The Transistor 2N2222 is used here for the simulation.

2) Principle of Operation

For a BJT to work as an amplifier, it must be first ensured that the BJT is operating in the Active region. For the transistor to be in active region the base to emitter junction should be in forward bias and the collector to base junction should be in reverse bias, if this condition is violated the transistor will not work as an amplifier. To ensure transistor is in active region, the resistances R1, R2, R3 and R4 are chosen such that they keep the transistor in active region i.e. base to emitter junction forward bias and collector to base junction reverse bias.

As shown in Fig 10.1, capacitor C1 is the Emitter bypass capacitor, C1 is used in parallel with R4 to provide a low reactance path to the amplified AC signal. If the Emitter bypass capacitor is not used, then the amplified AC signal passes through R4 causing a voltage drop across it, thereby reducing the output voltage.

The capacitor C2 is used to couple the signal to the base of the transistor. If this coupling capacitor is not used, the signal source resistance, rs will come across R2 and thus change the bias. C2 allows only AC signal to flow but isolates the signal source from R2.

The coupling capacitor, C3 couples one stage of amplification to the next stage. If this capacitor is not used, the bias conditions of the next stage will be drastically changed due to the shunting effect of R3. This is because R3 will come in parallel with the upper resistance R1 of the biasing network of the next stage, thereby altering the biasing conditions of the latter. In short, the coupling capacitor C3 isolates the dc of one stage from the next stage, but allows the Ac signal to pass.

Note: For the value of all the resistances used here, one can use any text book on BJT biasing. For the values of all the capacitors one can chose 1uF. Our main focus here is the simulation and the analysis of the simulation results.

3) Operating Point Analysis

The DC operating point analysis gives the voltages and currents at each node in the circuit.

Syntax for DC operating point analysis: .op

Figure 10.2 is shows the result of DC operating point analysis. From figure 10.2 it is clear that the base to emitter junction is forward biased and collector to base junction is reverse biased. So the transistor is operating in the active region. Kindly note that this is just an example circuit

Electronics Circuit SPICE Simulations with LTspice

and we have already ensured that transistor is in the active region. If the transistor is not in the active region one has to change the input voltage repeatedly to get the desired result. Now one can go for further analysis. Please note that the figure 10.1 shows just the circuit diagram and not the actual values of bias voltages. For the DC analysis a different DC source is used in place of the above mentioned source as shown in figure 10.1.

Note: During DC analysis the simulator accept only DC signal and ignores rest signals.

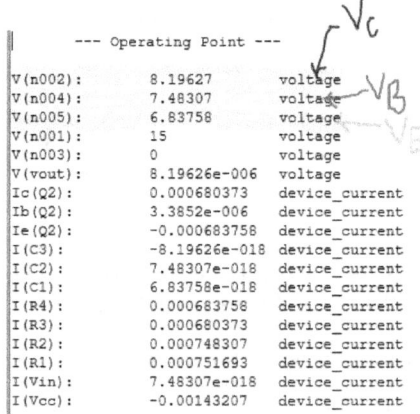

```
        --- Operating Point ---

V(n002):       8.19627        voltage
V(n004):       7.48307        voltage
V(n005):       6.83758        voltage
V(n001):       15             voltage
V(n003):       0              voltage
V(vout):       8.19626e-006   voltage
Ic(Q2):        0.000680373    device_current
Ib(Q2):        3.3852e-006    device_current
Ie(Q2):        -0.000683758   device_current
I(C3):         -8.19626e-018  device_current
I(C2):         7.48307e-018   device_current
I(C1):         6.83758e-018   device_current
I(R4):         0.000683758    device_current
I(R3):         0.000680373    device_current
I(R2):         0.000748307    device_current
I(R1):         0.000751693    device_current
I(Vin):        7.48307e-018   device_current
I(Vcc):        -0.00143207    device_current
```

Fig 10.3: Result of operating point analysis

4) AC Analysis

The AC analysis gives the frequency response of the circuit. For the AC analysis there must be one AC source in the circuit and for the simplicity in the calculation one must put the magnitude of AC signal as 1V. The AC source is superimposed on the input signal. Please note that the AC signal is superimposed on the DC bias voltage which is calculated from the above DC analysis. One can connect DC and AC sources in series across the base and ground for the input purpose.

Common Emitter Amplifier

Syntax for AC analysis: .ac dec 100 1 10G

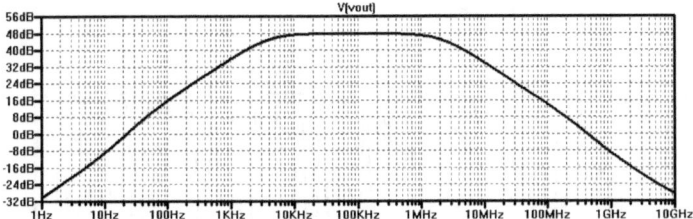

Figure 10.4: Result of AC analysis

From figure 10.3 it is clear that the gain is maximum between 5kHz to 5MHz frequency, beyond this frequency gain degrades. The voltage gain of an amplifier differs with frequency of the input signal. This is because the reactance of the capacitors in the amplifier circuit changes with the frequency of the input signal and thus the output is affected.

5) Transient Analysis

In transient analysis the SPICE simulator calculates the voltages and currents at each node at each time stamp. This analysis is shown here to demonstrate the amplification action of the common emitter amplifier. For the transient analysis also one must superimpose any sinusoidal or pulse signal on the DC bias voltage which is calculated from the above analysis otherwise no amplification action will be observed.

Syntax for transient analysis: .tran 0 500u 200u 0.01u

Electronics Circuit SPICE Simulations with LTspice

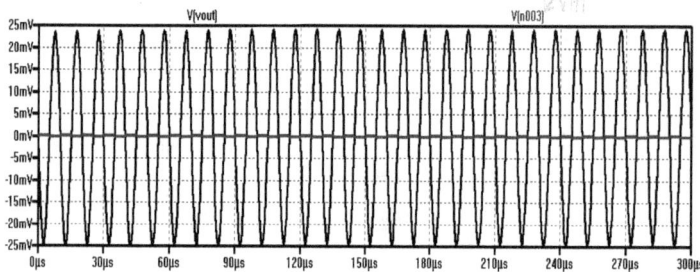

Fig 10.4: Result of transient analysis (voltage amplification action)

Figure 10.4 shows the simulation results of the transient analysis (voltage amplification action). From figure 10.4 it can be observed that input voltage is 0.2mV pp and output voltage is 50mV pp. So the voltage gain is **250.**

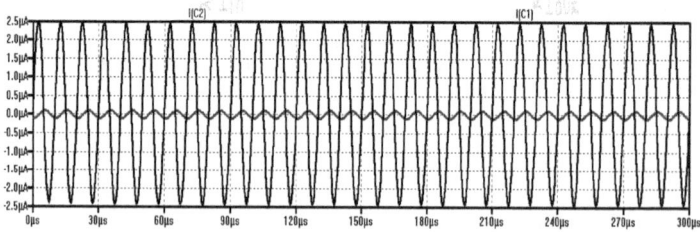

Fig 10.5: Result of transient analysis (current amplification action)

Figure 5.5 is showing the transient analysis (current amplification action). From figure 10.5 one can get that input current is 200nA pp and output current is 5µA pp. So the current gain is **25.**

Power gain is, voltage gain multiplied by the current gain that is **6250.**

Thus we can infer that a common emitter amplifier gives a high voltage gain, medium current gain and very high power gain.

****

Chapter 11   Metal Oxide Field Effect Transistor (MOSFET)

The concept of a MOSFET

MOSFET stands for Metal Oxide Semiconductor Field Effect Transistor (M= Metal, O= Oxide, S= Semiconductor, F= Field, E= Effect, T= Transistor). The first three terms Metal, Oxide and semiconductor define the structure of the transistor in the same order. The next two terms Field and Effect mark the operation of the device. A MOSFET works on the principle of the effect of Electric Field. Effectively an Electric field causes the creation or depletion of a channel in the device through which a current can flow.

Definition:

MOSFET is a three terminal semiconductor device that can work like either an amplifier or a switch based on the region of its operation. There is also a fourth terminal called Substrate that may or may not be available. In case of a NMOS we generally connect the substrate terminal it to the ground terminal and in case of a PMOS we connect it to the supply voltage(VDD). The four terminals are Source, Gate, Drain and Substrate as shown in Fig 11.1.

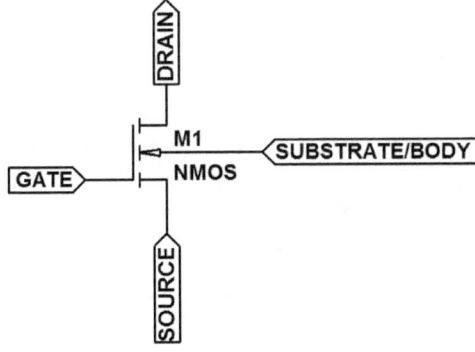

Fig 11.1: An N-Channel MOSFET(NMOS4 in LTspice)

A MOSFET operates in three regions.

1) Linear Region
2) Saturation Region
3) Cutoff Region

V-I Characteristics of a N channel Enhancement MOSFET

1) Drain Characteristics: Drain Current (I_D) is measured by varying Drain to source voltage (V_{DS}) by keeping the gate to source Voltage (V_{GS}) fixed.

2) Transfer Characteristics: Drain Current (I_D) is measured by varying Gate to Source Voltage (V_{GS}) and keeping Drain to Source voltage (V_{DS}) fixed such that the MOSFET is in Saturation region. For transfer characteristics it is ensured that

$$V_{DS} > V_{DSSAT} = V_{GS} - V_T$$

One gets Identical waveforms for all values of $V_{DS} \geq V_{DSSAT}$.

MOSFET Declaration

A MOSFET in the netlist is declared as:

MOFET_Name ND NG NS NB MODEL_Name

There are other parameters that are optional, namely: Length (L), Width (W), Drain Area (AD), Source Area (AS), Drain Perimeter (PD), Source Perimeter (PS), No. Parallel Devices (M). If these parameters are not set the LTspice Simulator uses defaults values for Simulation.

The default values of W and L values are 100μm.

W = defl = 100μm - Default MOS channel length

L= defw= 100μm - Default MOS channel width

Additional MOSFET parameters can be defined in MOSFET Model file. In LTspice if you don't specify any parameter value the default level is one.

The default model file is similar to the one shown below with some more SPICE parameters.

.model NMOS4 NMOS (LEVEL=1 VTO=0 KP= 20u GAMMA=0 LAMBDA=0 TOX=1E-7)

We can add a model file to the Schematic using .include Spice directive. But for our present simulation we will go with the default values only.

Note that the simulations in this chapter are based on default values as a result they may vary slightly with the standard results. Nevertheless a better picture can be seen by adding a MOSFET model file to the schematic.

1) Drain Characteristics

In order to plot the Drain Characteristics draw the Schematic Diagram as shown in Fig 11.5. Start with selecting NMOS4 from the component window as shown in Fig 11.2. Then right click on Symbol to get the Optional Parameter Declaration Window as shown in Fig 11.4.

*Metal Oxide Field Effect Transistor
(MOSFET)*

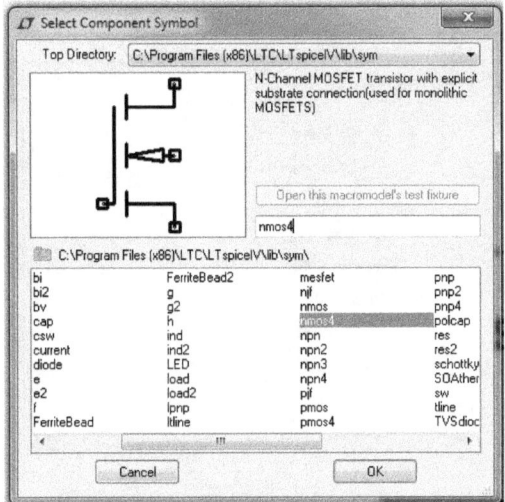

Fig 11.2: Choosing NMOS4 from the component window

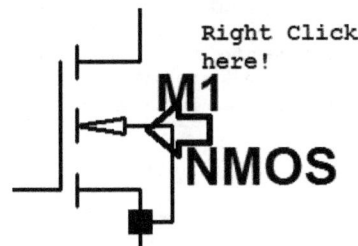

Fig 11.3: Getting the Device Declaration Window- NMOS4

Electronics Circuit SPICE Simulations with LTspice

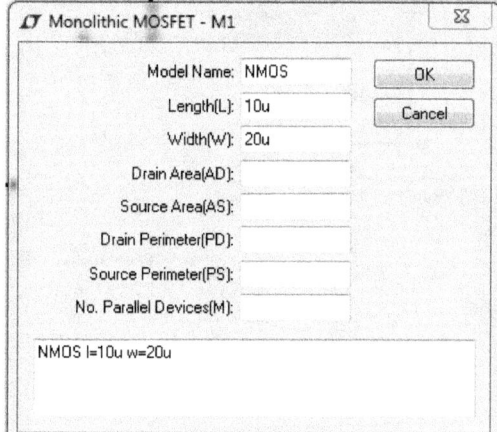

Fig 11.4: Device Declaration Window- NMOS4

Voltage Vs Current characteristics of a MOSFET gives insights into the behavior of the device when a biasing voltage is applied. For Drain characteristics, connect the circuit components as shown in Fig 11.5 and choose DC sweep from the Edit Simulation Command as shown in Fig 11.6. and Fig 11.7 Note that both V_{DS} and V_{GS} are being varied simultaneously. Since V_{DS} has to be along the x axis it is the 1st Source. V_{GS} is the second source and is kept fixed for different discrete values.

Metal Oxide Field Effect Transistor (MOSFET)

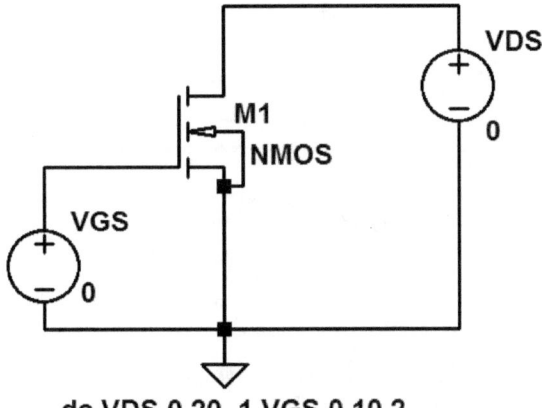

.dc VDS 0 20 .1 VGS 0 10 2

Fig 11.5: Schematic for Drain Characteristics

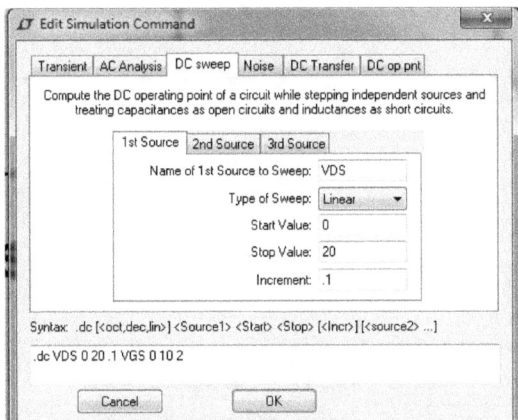

Fig 11.6: Setting the first Voltage source sweep values

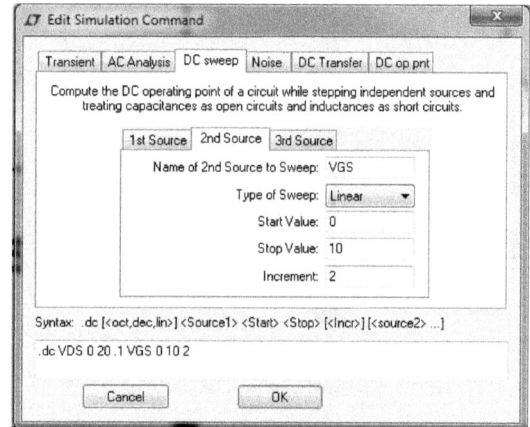

Fig 11.7: Setting the second Voltage source sweep values

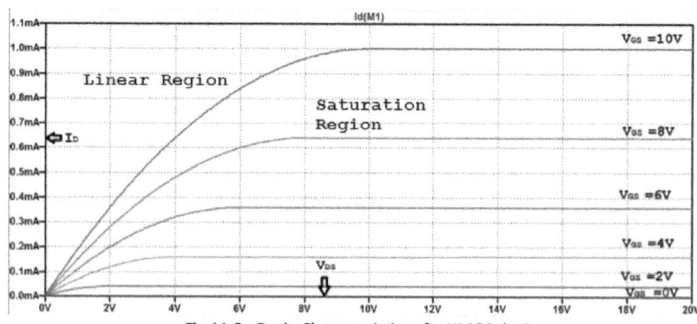

Fig 11.8: Drain Characteristics of a NMOS device

From Fig 11.8 we can observe that I_D increases linearly for lower values of VDS and then becomes almost constant for higher values of V_{DS} for a given fixed value VGS. Also for a fixed difference of 2 Volts in VGS values there is larger rise in the current I_D.

Metal Oxide Field Effect Transistor
(MOSFET)

2) Transfer Characteristics

In order to plot the Transfer Characteristics draw the Schematic Diagram as shown in Fig 11.9. Choose DC sweep from the Edit Simulation Command as shown in Fig 11.10. From Fig 11.11 we can observe that I_D rises sharply with increasing values of V_{GS}.

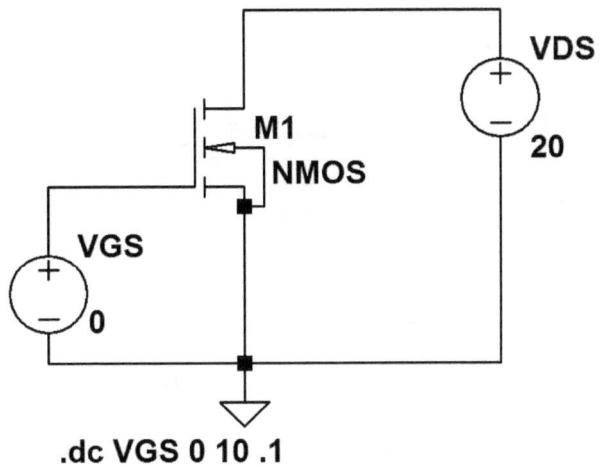

.dc VGS 0 10 .1

Fig 11.9: Schematic for Transfer Characteristics

Electronics Circuit SPICE Simulations with LTspice

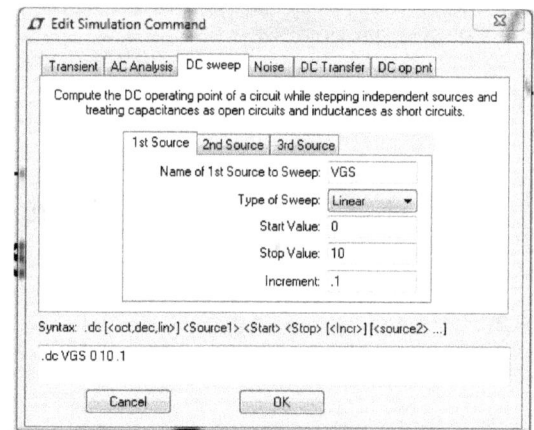

Fig 11.10: Setting the Voltage source-VGS sweep values

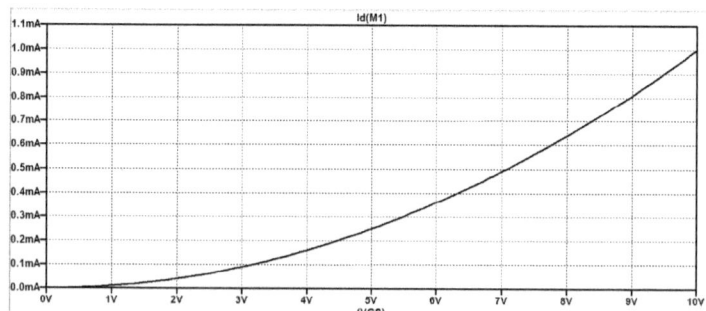

Fig 11.11: Transfer Characteristics- Typical Drain current Vs VGS plot for a NMOS device

****

CHAPTER-12   NOT Gate using a MOSFET

About the chapter

This chapter shows you the simulations of a NOT Gate using a MOSFET and a resistor. A NOT Gate can be designed using a MOSFET and a resistor.

NOT Gate

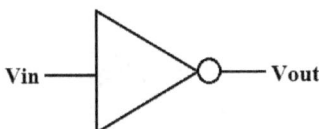

Fig 12.1: Symbol – NOT gate

Vin	Vout
0	1
1	0

Table 12.1: Truth Table - NOT gate

In a NOT gate, if the input is low (0V) then the output is high (5V) and vice versa. The truth table of NOT gate can be seen from Table 12.1. In Table 12.1 we have the input as V1 (Vin) and the output is Vout which is measured at the drain of the MOSFET.

Operation of a NOT Gate using a Power MOSFET

Case 1: V1 = 5V

If V1 is high(5), the transistor M1 is turned ON. As a result a current starts to flow between the drain and the source terminals of the transistor. This leads to a drain current(ID) flowing through the resistor R1. Thus a drop of 5V occurs across the resistor R1 and the output voltage Vout becomes low(0). Fig 12.2 shows the schematic diagram of NOT Gate using MOSFET. The MOSFET used for the this simulation is BSB012N03LX3. As shown in Fig 12.3, this MOSFET can be selected by right clicking on the MOSFET symbol after placing it on the circuit schematic. Fig 12.4 shows Transient Analysis of the NOT Gate where we can clearly see that the output is constant at a value of about 0V.

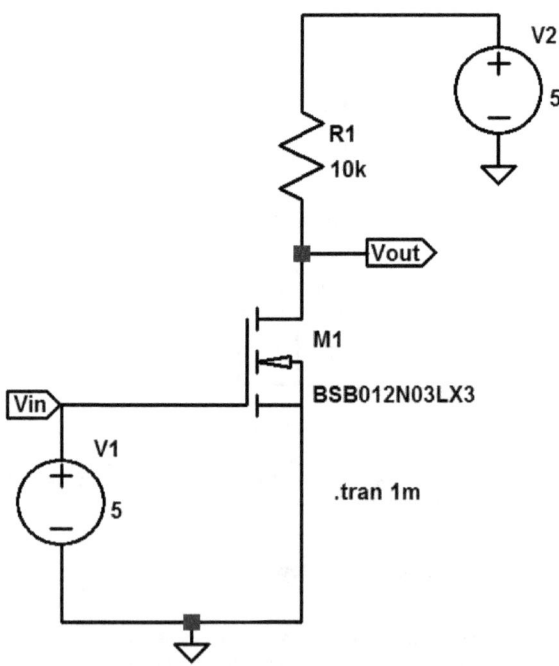

Fig 12.2: Schematic Diagram – NOT Gate using MOSFET when the input is 5V

Electronics Circuit SPICE Simulations with LTspice

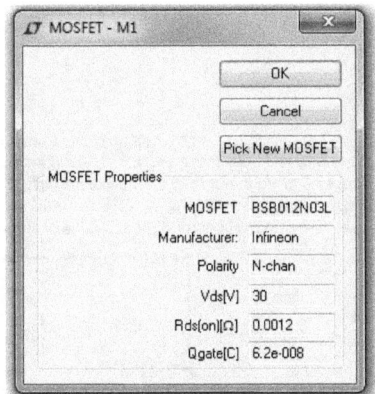

Fig 12.3 Choosing MOSFET BSB012N03LX3

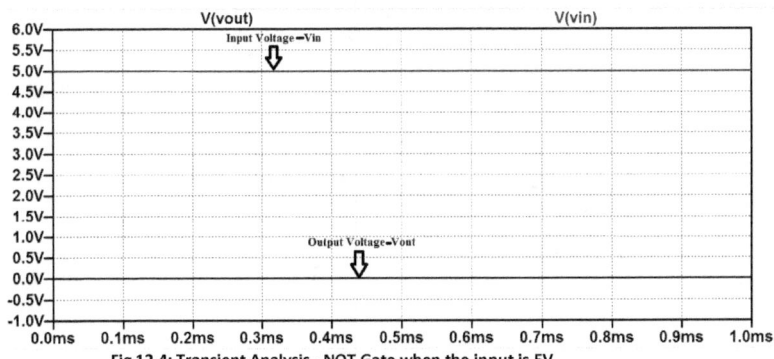

Fig 12.4: Transient Analysis - NOT Gate when the input is 5V

Case 2: V1= 0V

If V1 is low(0), the transistor M1 is turned OFF. As a result no current flows between the source and the drain terminals of the MOSFET. As a result there is no voltage drop across the resistor R1. The entire voltage V2 appears at Vout and thus Vout becomes high(5). Fig 12.5 shows the schematic diagram of NOT Gate using a MOSFET. Fig 12.6 shows Transient Analysis of the NOT Gate where we can clearly see that the output is constant at a value of about 5V.

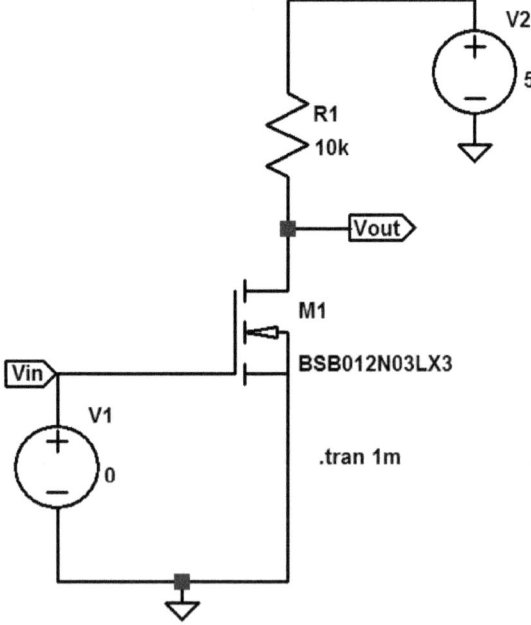

Fig 12.5: Schematic Diagram – NOT Gate using NPN Transistor when the input is 0V

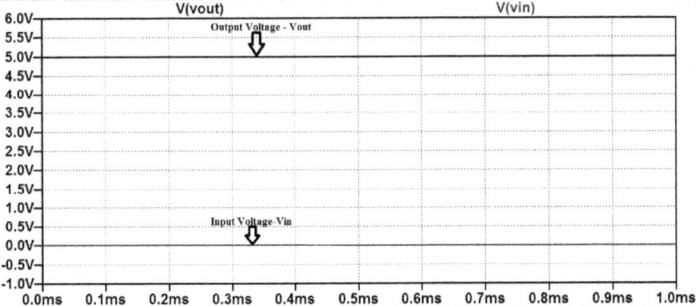

Fig 12.6: Transient Analysis - NOT Gate when the input is 0V

DC Sweep Analysis

For the DC Sweep Analysis, choose DC sweep from the Edit Simulation Command and set the Voltage source V1 as shown in Fig 12.7. From DC Sweep Analysis we can observe that when the input voltage is low, the output voltage is high and vice versa.

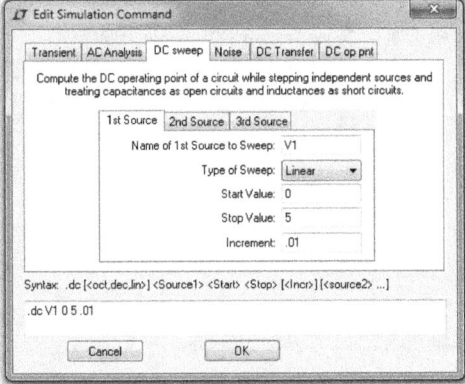

Fig 12.7: Setting the Input Voltage source sweep values

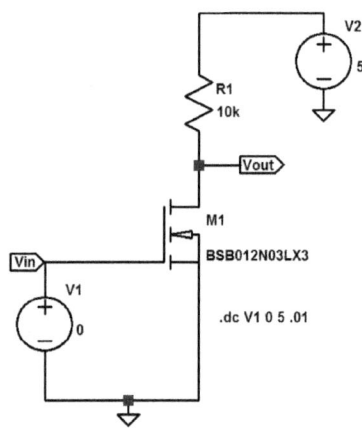

Fig 12.8: DC Sweep Analysis- NOT Gate using MOSFET

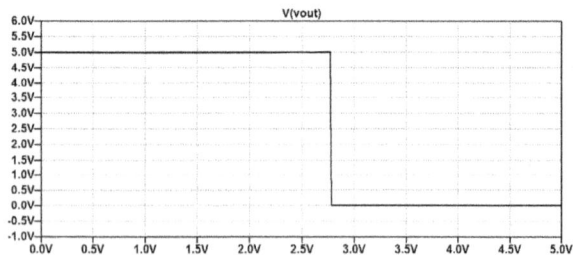

Fig 12.9: DC Sweep Analysis- Vout Vs Vin

****

Chapter 13   Common Source Amplifier Using MOSFET

This chapter shows you the simulations of the Common Source Amplifier Using MOSFET, its biasing, AC analysis and the gain.

Contents of the chapter:
1) Circuit diagram
2) Principle of Operation
3) Operating point selection
4) Transient analysis
5) Calculation of output resistance

1) Circuit diagram

To design the common source amplifier we have to choose a NMOS transistor and assign channel length and channel width to the transistor. To assign channel length and channel width one has to right click on the icon of NMOS, then a window will open as shown in figure 13.1, assign the channel length and width. Connect resistance R1 between drain of NMOS and VDD. The output is taken from drain of NMOS. The capacitor C1 is connected at the drain of NMOS to filter out DC component and pass AC signal so that during transient analysis one can measure only the AC output to identify the gain of the amplifier.

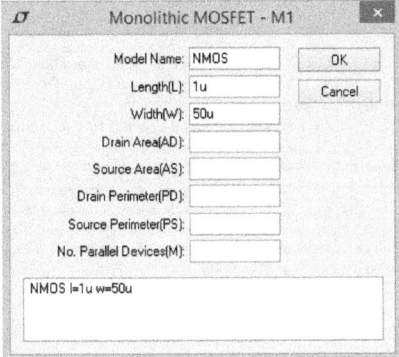

Fig 13.1: Window to assign channel length and channel width to NMOS

Common Source Amplifier Using MOSFET

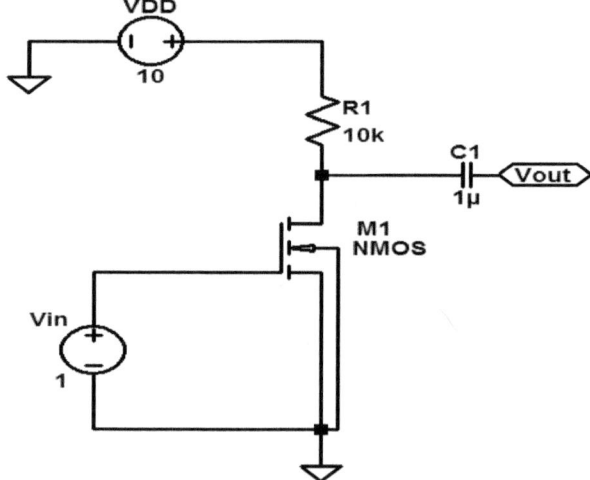

Figure 13.2: Circuit diagram of common source amplifier

2) Operating principle of Operation

To operate the transistor in the amplification mode one has to bias the transistor in the saturation mode. To bias the transistor in the saturation mode one has to give gate to source voltage (V_{gs}) greater than threshold voltage of the transistor. Once the transistor is biased in saturation mode one has to find at which input voltage the amplifier show maximum gain. To find the input voltage for the amplifier to show maximum gain, the procedure is explained in the next section. Once the input voltage is fixed a small AC voltage is superimposed on the DC input voltage and transient analysis will be done, the amplified AC output voltage can be measured at Vout node.

3) Operating point selection

To find the operating point one has to DC sweep the input voltage from 0V to VDD and plot the output voltage with respect to input voltage as shown in the figure 13.3. The output voltage is measured at the drain of NMOS.

Electronics Circuit SPICE Simulations with LTspice

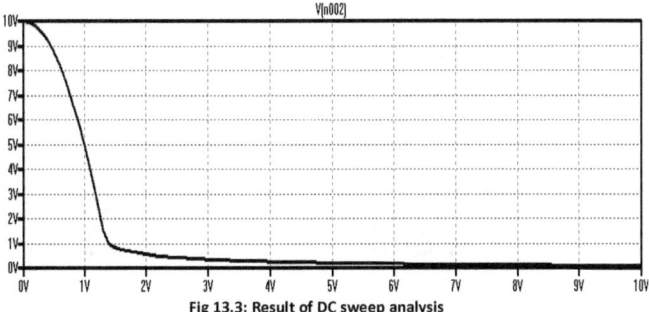

Fig 13.3: Result of DC sweep analysis

Syntax for DC sweep: .dc Vin 0 10 0.1

Now one has to differentiate the output voltage with respect to input voltage as shown in figure 13.4, the figure 13.4 also shows you the command to differentiate Y-axis with respect to X-axis. Select the operating point for the desired gain.

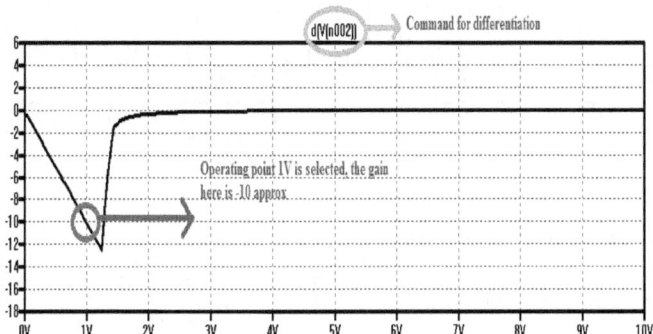

Fig 13.4: Showing the differentiation of output with respect to input voltage

Common Source Amplifier Using MOSFET

4) Transient analysis

Figure 13.5 is shows the transient analysis of the Common Source Amplifier. Make sure for the transient analysis the sinusoidal signal is superimposed on the DC bias voltage otherwise the amplification action will not be observed. From Figure 13.5 the amplification action of the amplifier can be observed. From the figure 13.5 it is clear that the gain of the amplifier is -10.

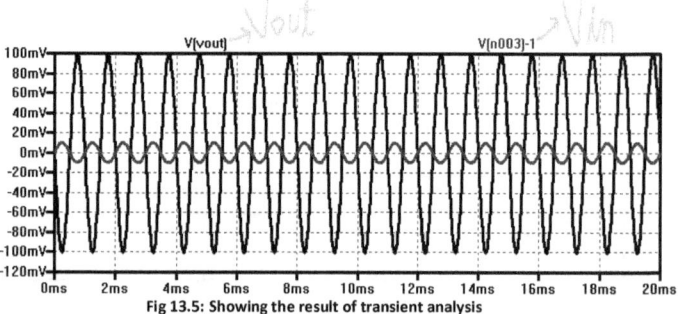

Fig 13.5: Showing the result of transient analysis

Syntax for transient analysis: .tran 0 20m 0 0.01m

5) Calculation of output resistance

For the calculation of output resistance one has to find the DC voltage at the node A (output node shown in figure 13.6). Then connect a separate voltage source parallel to the output node and set the DC voltage of that source equal to the DC voltage at node A and also set AC voltage to be 1V for the simplicity. Now do the AC analysis and measure the current across that voltage source. The output resistance is AC voltage divided by AC current. Figure 13.6 shows the setup for the calculation of output resistance.

Syntax for AC analysis: .ac dec 100 10 100G

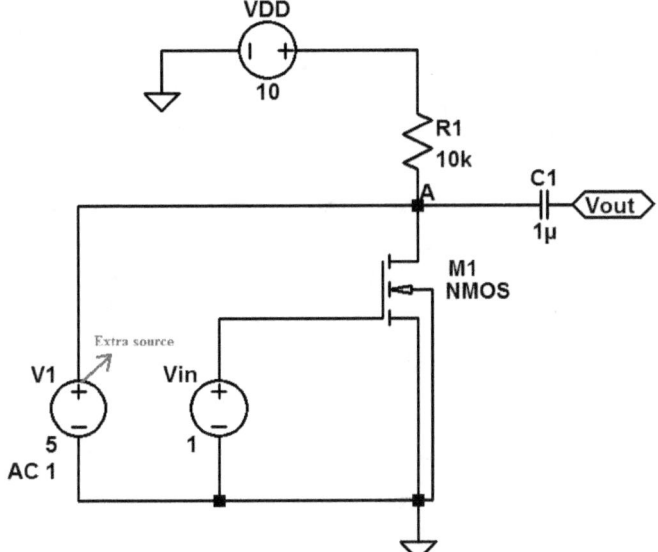

Fig 13.6: setup for the calculation of output resistance

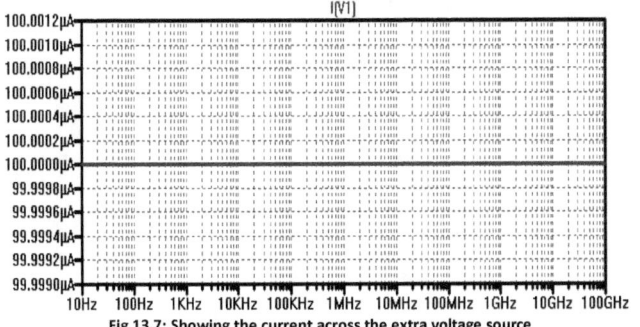

Fig 13.7: Showing the current across the extra voltage source

Common Source Amplifier Using MOSFET

In order to convert Y-axis of figure 13.7 into linear scale from dB scale, left click on the Y-axis, a window will open as shown in figure 13.8. Select linear Representation from this window.

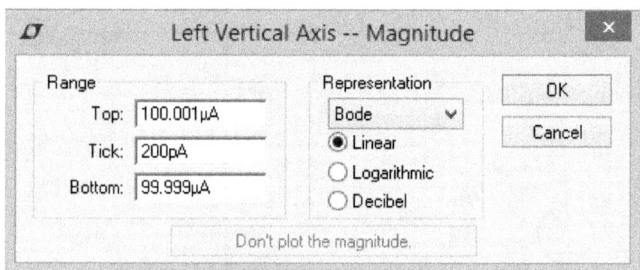

Fig 13.8: Window for changing the content of Y-axis into linear or dB

From the figure 13.7, the value of AC current is 100 µA so the output resistance is **10 KΩ**.

****

Chapter-14   Operational Amplifier (OpAmp)

The concept of an Operational Amplifier (OpAmp)

Definition: An Operational is a five terminal semiconductor device that allows one to perform mathematical operations such as addition, subtraction, integration and differentiation on analog signals. These terminals are Inverting Voltage Input, Non-Inverting Voltage Input, Voltage output, Positive power supply input and Negative power supply input. The Fig 14.1 below shows a typical OpAmp-LT6016 from Linear Technology. There are many different OpAmps available in LTspice with different features that one can use for designing a circuit. You can typically find these OpAmps by navigating through the following path up to the folder Opamps, C:\Program Files (x86)\LTC\LTspiceIV\lib\sym\Opamps, for example- LM308, LT1006, LT1112, LTC6078 and many more.

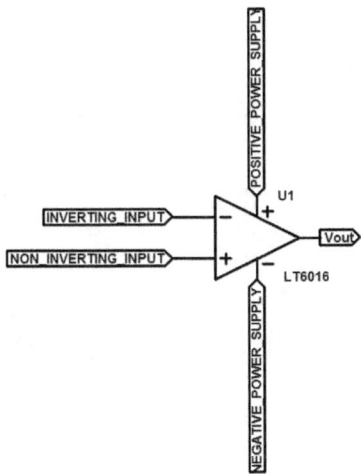

Fig 14.1: A Typical OpAmp-LT6016 from Linear Technology

However in all our simulations in this book we will be using UniversalOpamp2 which is available in LTspice for educational purposes. UniversalOpamp2 can be selected from the component window as shown in Fig 14.2.

Operational Amplifier (OpAmp)

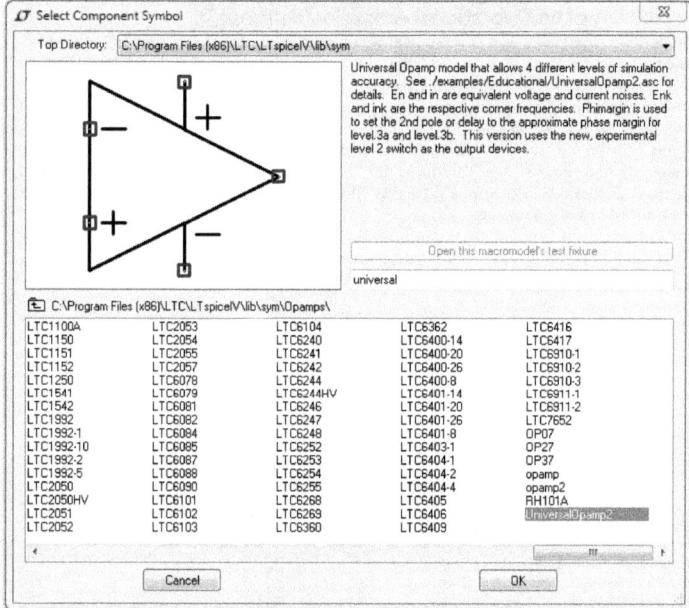

Fig 14.2: selecting UniversalOpamp2

OpAmp Configurations

OpAmp can operate in two different configurations.
a) Open loop configuration
b) Closed loop configuration

a) Open loop configuration: In this configuration the Vout depends only on the input signals. There is no feedback connection from the output to the input of the OpAmp.

Electronics Circuit SPICE Simulations with LTspice

Open loop gain (A_{OL}): Open loop gain is the Voltage gain of an OpAmp when it is in the Open loop configuration. A_{OL} is a parameter that depends on the frequency. At higher frequencies A_{OL} falls off rapidly. However it is at its highest values at 0Hz or DC.

$$V_{out} = A_{OL} (V_{noninverting} - V_{inverting})$$
14.1

When in Open loop Configuration an OpAmp can work in following three input modes:

1) Differential Input Mode: The input Signal is applied to both the input terminals (Inverting and Non Inverting). If both these signals are 180 degree out of phase with each other then the output is completely in phase with the signal that is available at the non inverting input. If both the signals are in phase, i.e. if they are identical, then there is no output (0V), these inputs are called common mode signals

Simulation with DC inputs

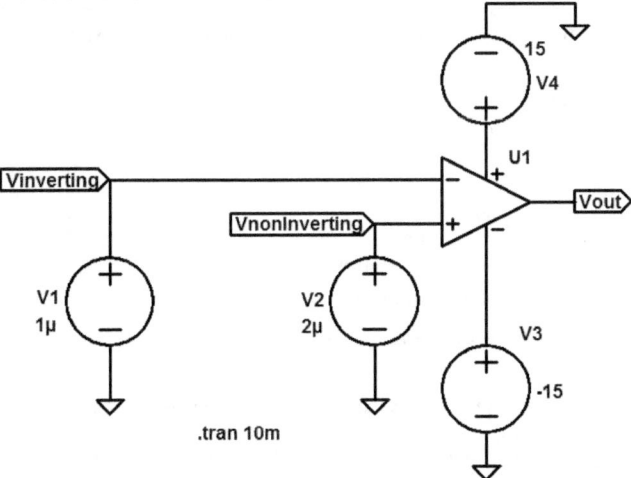

Fig 14.3: OpAmp in Differential Input Mode with DC Inputs(UniversalOpamp2 is used)

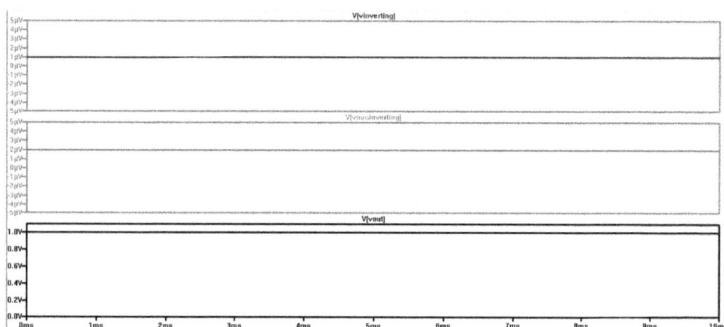

Fig 14.4: Transient Analysis plots - OpAmp in Differential Input Mode with DC Inputs

From equation 14.1:

$V_{out} = A_{OL}(V_{noninverting} - V_{inverting})$

From the OpAmp circuit in Fig 14.3 we have,

$V_{noninverting} = 2\mu V$

$V_{inverting} = 1\mu V$

$V_{out} = A_{OL}(2\mu V - 1\mu V) = A_{OL}1\mu V$

But from the simulation results in Fig 14.4 we have, $V_{out} = 1V$.

Substituting the value of V_{out} in the above equation we have, $A_{OL} = \dfrac{1V}{1\mu V} = 10^6$.

Thus the UniversalOpamp2 has a open loop gain(DC) $= 10^6$.

Simulation with Out of phase Sinusoidal inputs having frequency 10Hz

Inverting Input has been set to an amplitude of $5\mu V$ and frequency of 10Hz and non-Inverting Input has also been set to an amplitude $5\mu V$ and frequency of 10Hz but with a phase shift of 180° as shown in Fig 14.5.

Electronics Circuit SPICE Simulations with LTspice

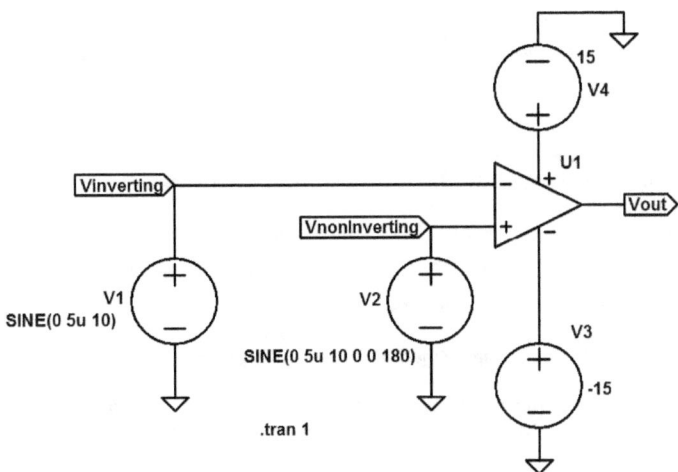

Fig 14.5: OpAmp in Differential Input Mode with out of phase Sinusoidal Inputs having frequency 10Hz

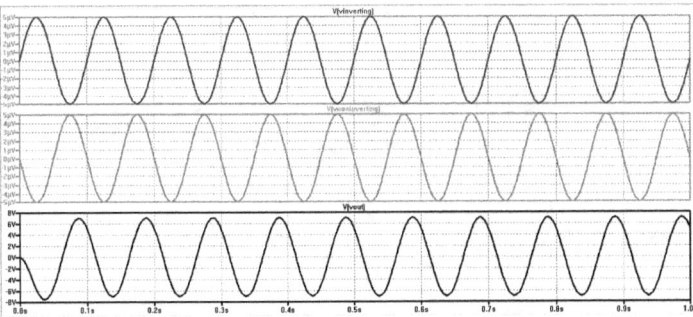

Fig 14.6: Transient Analysis plots - OpAmp in Differential Input Mode with out of phase Sinusoidal Inputs having frequency 10Hz

Operational Amplifier (OpAmp)

From equation 14.1:

$V_{out} = A_{OL} (V_{noninverting} - V_{inverting})$

From the OpAmp circuit in Fig 14.5, the maximum difference between the input signals is $10\mu V$.

$V_{out} = A_{OL} 10\mu V$

But from the simulation results in Fig 14.6 we have the peak value of $V_{out} = 7V$.

Substituting the value of V_{out} in the above equation we have, $A_{OL} = \dfrac{7V}{10\mu V} = 0.7 \times 10^6$.

This shows that the A_{OL} value has reduced to some extent at 10 Hz.

Simulation with Out of phase Sinusoidal inputs having frequency 50Hz

Inverting Input has been set to an amplitude of $10\mu V$ and frequency of 10Hz and nonInverting Input has also been set to an amplitude $10\mu V$ and frequency of 10Hz but with a phase shift of $180°$ as shown Fig 14.7.

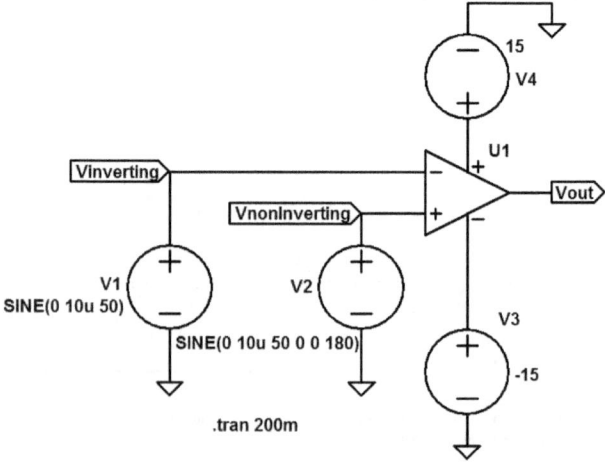

Fig 14.7: OpAmp in Differential Input Mode with out of phase Sinusoidal Inputs having frequency 50Hz

Electronics Circuit SPICE Simulations with LTspice

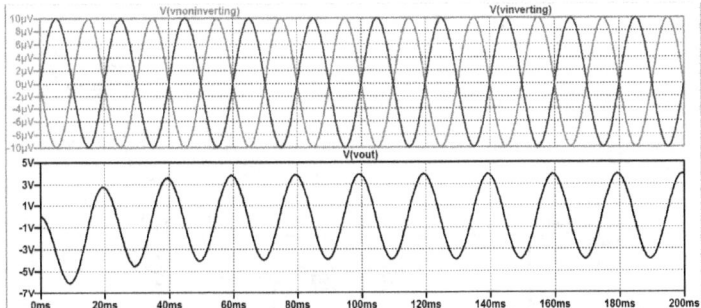

Fig 14.8: Transient Analysis plots - OpAmp in Differential Input Mode with out of phase Sinusoidal Inputs having frequency 50Hz

From equation 14.1:

$V_{out} = A_{OL}(V_{noninverting} - V_{inverting})$

From the OpAmp circuit in Fig 14.7 the maximum difference between the input signals is, $20\mu V$.

$V_{out} = A_{OL} 20\mu V$

But from the simulation results in Fig 14.8 we have the peak value of $V_{out} = 4V$.

Substituting the value of V_{out} in the above equation we have, $A_{OL} = \dfrac{4V}{20\mu V} = 0.2 \times 10^6$.

This shows that the A_{OL} value has reduced considerably at 50 Hz.

Simulation with common mode Sinusoidal inputs

In this case both the input signals are in phase as shown in fig 14.9. From the Transient Analysis plots in Fig 14.10 we can see that Vout =0.

Operational Amplifier (OpAmp)

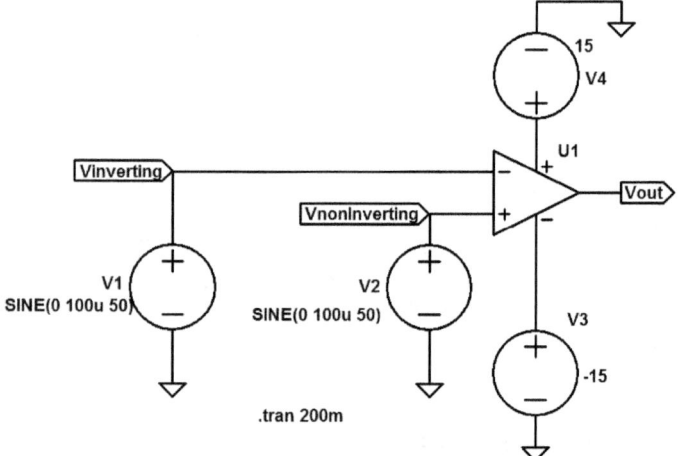

Fig 14.9: OpAmp in Differential Input Mode with common mode inputs having frequency 50Hz

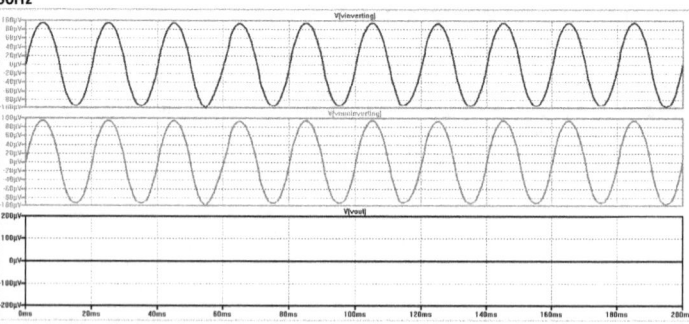

Fig 14.10: Transient Analysis plots - OpAmp in Differential Input Mode with common mode inputs having frequency 50Hz

Electronics Circuit SPICE Simulations with LTspice

2) Non-Inverting Input Mode: The input signal is applied only to Non-Inverting terminal and the Inverting terminal is grounded. The output Signal in this case is completely in phase with the signal that is available at the non-inverting input. Sinusoidal input Signal of frequency 10Hz has been applied to the Non Inverting input terminal as shown in Fig 14.11. From the Transient Analysis plots in Fig 14.12 we can see that the output Signal is completely in phase with the signal that is available at the non-inverting input. Also the output Signal is 0.7×10^6 times the input signal.

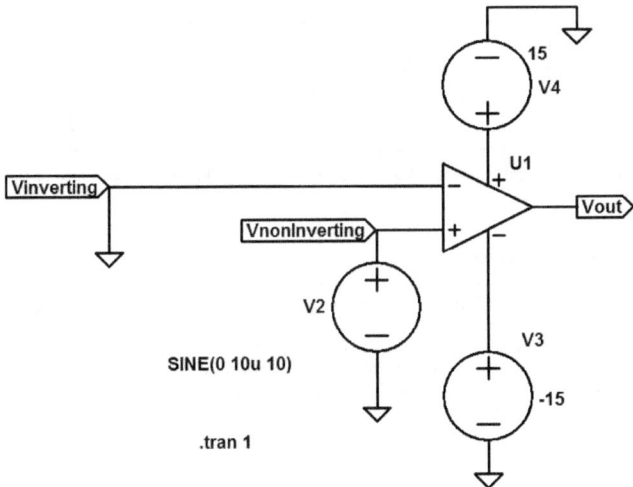

Fig 14.11: OpAmp in Non-Inverting Input Mode having sinusoidal input of frequency 10Hz

Operational Amplifier (OpAmp)

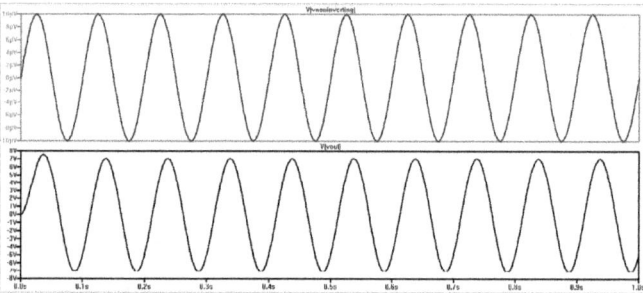

Fig 14.12: Transient Analysis plots - OpAmp in Non-Inverting Input Mode having sinusoidal input of frequency 10Hz

3) Inverting Input Mode

The input Signal is applied only to Inverting terminal and the Non-Inverting terminal is grounded. The output Signal in this case is 180 degree out of phase with the signal that is available at the inverting input, thus the name inverting input. Sinusoidal input Signal of frequency 10Hz has been applied to the inverting input terminal as shown in Fig 14.13. From the Transient Analysis plots in Fig 14.14 we can see that the output Signal is 180 degree out of phase with the signal that is available at the inverting input. Also the output Signal is 0.7×10^6 times the input signal.

Electronics Circuit SPICE Simulations with LTspice

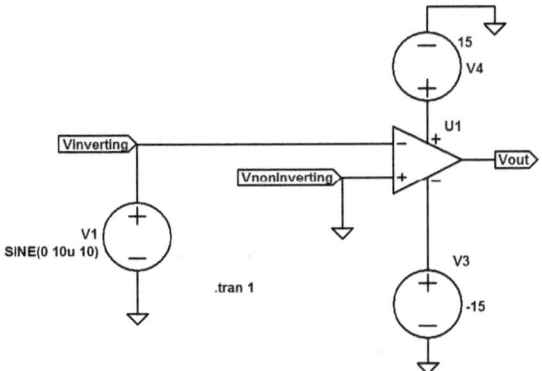

Fig 14.13: OpAmp in Inverting Input Mode having sinusoidal input of frequency 10Hz

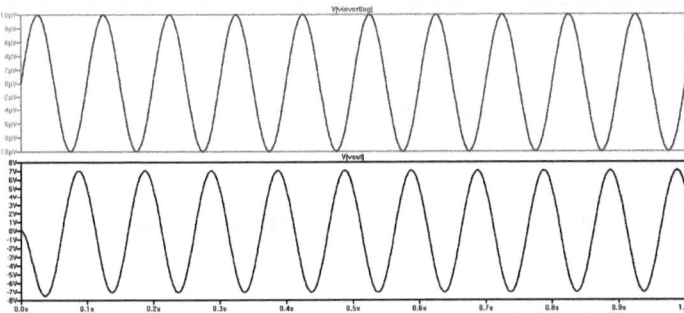

Fig 14.14 Transient Analysis plots - OpAmp in Inverting Input Mode having sinusoidal input of frequency 10Hz

Open loop AC Analysis (Frequency Response)

AC Analysis of the OpAmp in open loop configuration is performed to obtain a plot that shows different values of gain when the frequency of the input signal is varied. Such a plot is referred to as the Frequency Response plot.

Operational Amplifier (OpAmp)

In order to perform AC Analysis of the circuit, go to edit Simulation Command Window as shown in Figure 14.15 and choose AC Analysis. Four Types of Sweeps are possible, namely Decade, Octave, Linear and List. Here we have chosen Decade. The frequency range of sweep is 1 Hz to 10GHz. Fig 14.16 shows the circuit for AC Analysis with frequency range from 1 Hz to 10GHz while Fig 14.20 shows the same circuit for AC Analysis with frequency range from 1 Hz to 100GHz.

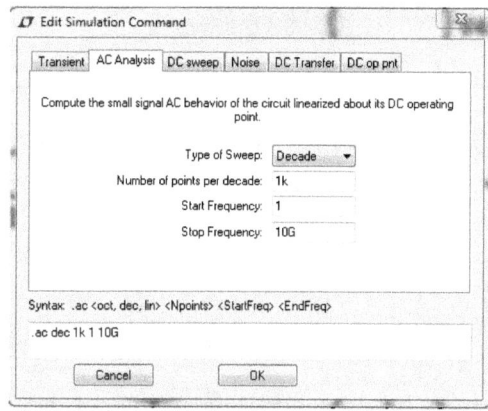

Fig 14.15: selecting AC Analysis from Edit Simulation Command

Electronics Circuit SPICE Simulations with LTspice

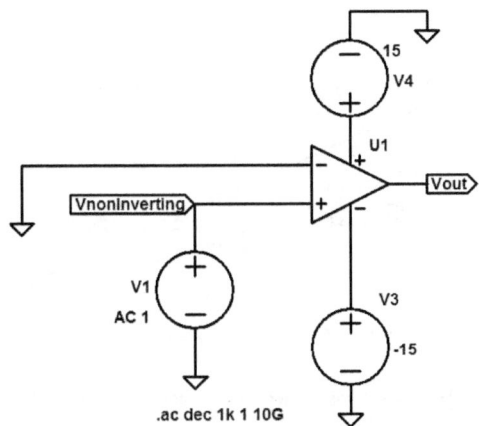

.ac dec 1k 1 10G

Fig 14.16: AC Analysis for frequency range 1 Hz to 10 GHz

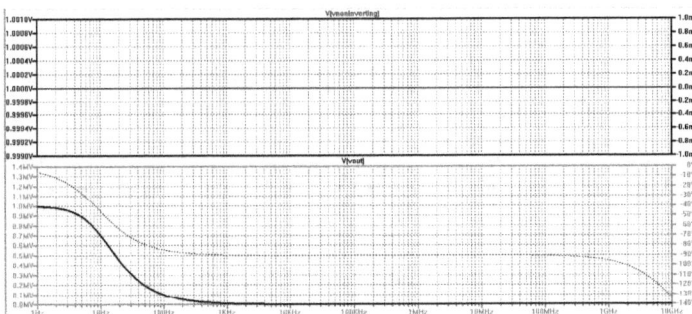

Fig 14.17: Non inverting input Signal and AC Analysis plot for frequency range 1 Hz to 10 GHz in linear scale

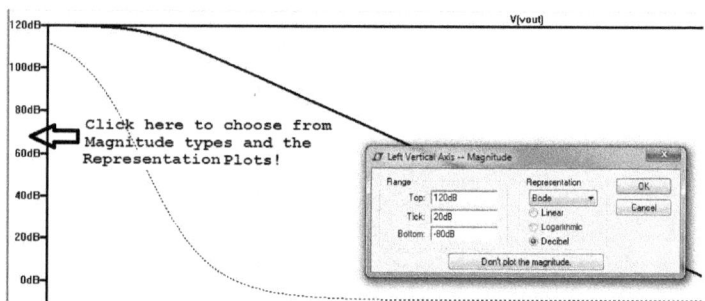

Fig 14.18: Choosing from Magnitude types and Representation Plots.

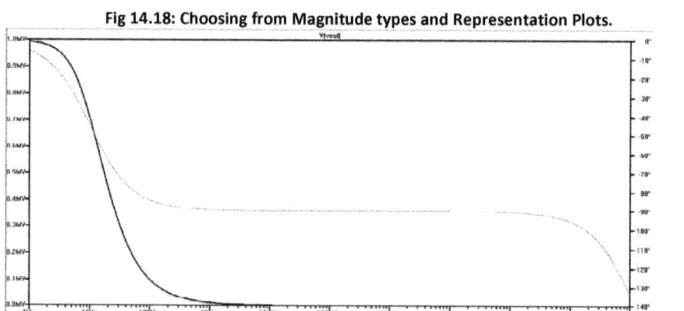

Fig 14.19: AC Analysis plot for frequency range 1 Hz to 10GHz in linear scale (Expanded view)

Electronics Circuit SPICE Simulations with LTspice

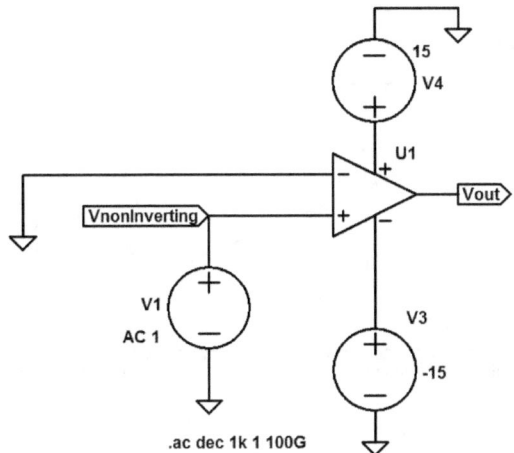

.ac dec 1k 1 100G

Fig 14.20: AC Analysis plot for frequency range 1 Hz to 100 GHz

Here the frequency range has been extended to 100GHz and the open loop gain has been plotted.

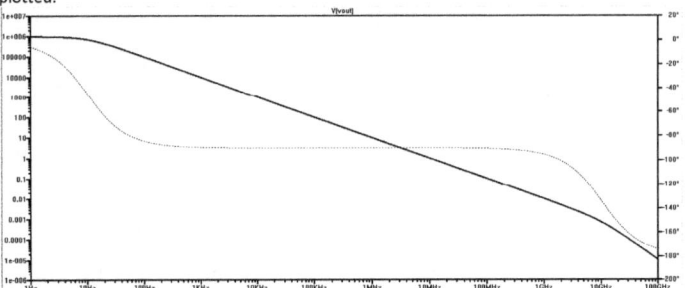

Fig 14.21: AC Analysis plot for frequency range 1 Hz to 100 GHz with magnitude representation in logarithmic scale

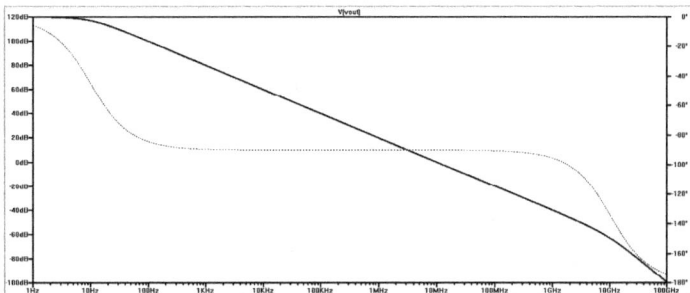

Fig 14.22: AC Analysis plot for frequency range 1 Hz to 100 GHz with magnitude representation in Decibel scale

b) Closed loop configuration: In the closed loop Configuration Vout is connected to the Inverting terminal of the OpAmp via a resistive path. Closed loop configuration has been discussed in detail in the next Chapter.

****

Chapter-15   Inverting Amplifier using OpAmp

As discussed earlier OpAmp can operate in two different configurations.
a) Open loop configuration
b) Closed loop configuration
Open loop Configuration has been discussed in detail in the last chapter.

OpAmp in Closed loop configuration

In the closed loop Configuration Vout is connected to the Input of the OpAmp via a resistive path. In other words there is a feedback path available in this case where the output is fed back to the input of the OpAmp. A huge majority of OpAmp circuits are used in closed loop configuration where there is a feedback path. In fact OpAmps are rarely used in open loop configuration. OpAmp is widely used in analog signal processing and there are innumerous applications of OpAmp as an amplifier.

Designing an amplifier with an OpAmp in closed loop configuration is the simplest and best practice. The Voltage gain can be set merely by changing the value of resistances in the feedback path and on the input side of the amplifier. As compared to the BJT and FET amplifiers where one has to properly bias the circuit, OpAmp amplifiers have no such complications and are much easier to handle.

OpAmp Amplifiers are mainly of two kinds:
a) Inverting Amplifier
b) Non- inverting Amplifier

In this chapter inverting amplifier has been discussed. The next chapter covers the non-inverting amplifier.

The concept of Inverting Amplifier using OpAmp

In an inverting amplifier, the input signal is applied to the inverting input of the OpAmp. The non-inverting input of the OpAmp is kept grounded. The Output terminal of the OpAmp is connected to the inverting input terminal via a resistor of suitable value. Fig 15.1 shows the circuit diagram for the inverting amplifier, R1 is the feedback resistor and R2 is the input resistor. A sinusoidal input of maximum voltage=5V has been applied at the inverting input of the amplifier circuit. Fig 15.2 shows the transient analysis plots for various Voltage levels and Fig 15.3 shows Transient Analysis plots for various current levels in the inverting amplifier circuit.

The most important property of an amplifier is its Voltage gain. The closed loop voltage gain of the inverting amplifier has been calculated in the next section.

Inverting Amplifier using OpAmp

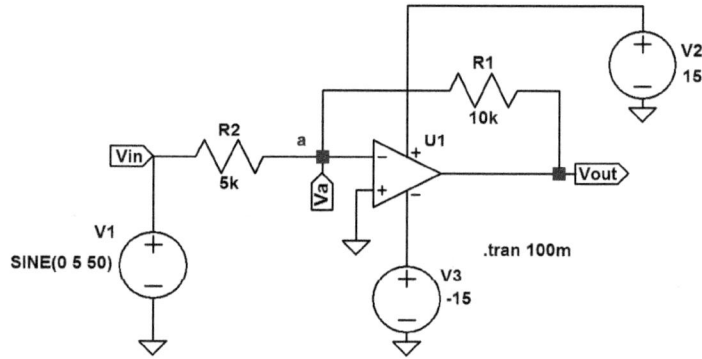

Fig 15.1: Circuit diagram: Inverting Amplifier

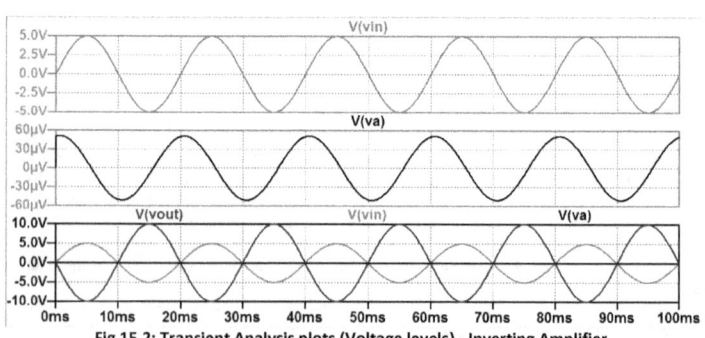

Fig 15.2: Transient Analysis plots (Voltage levels) - Inverting Amplifier

Electronics Circuit SPICE Simulations with LTspice

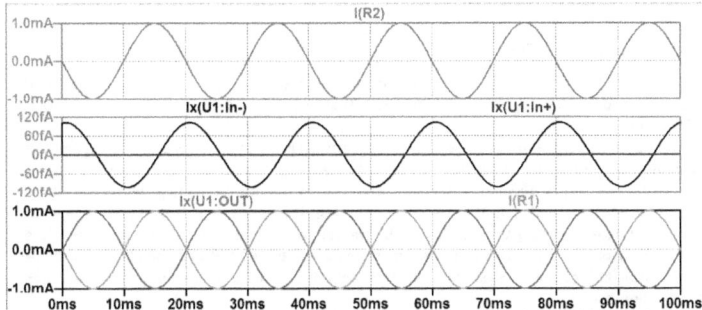

Fig 15.3: Transient Analysis plots (current levels) - Inverting Amplifier

Inverting amplifier-Closed Loop Voltage Gain

The gain of an inverting amplifier can be found out by applying nodal analysis at node a.

Applying Nodal Analysis at Node a, we have,

$$\frac{V_a - V_{in}}{R_2} + \frac{V_a - V_{out}}{R_1} = 0 \qquad 15.1$$

where V_a represents the voltage at node a.

Now, since V_a is at the same potential as the ground(Virtual ground).

$$V_a = 0$$

Substituting the value of V_a in 15.1 we get,

$$\frac{-V_{in}}{R_2} + \frac{-V_{out}}{R_1} = 0$$

$$\Rightarrow \frac{V_{out}}{R_1} = \frac{-V_{in}}{R_2}$$

$$\text{or,} \quad \frac{V_{out}}{V_{in}} = \frac{-R_1}{R_2}$$

The closed loop gain A_{CL} of an amplifier is given by :

$$A_{CL} = \frac{V_{out}}{V_{in}} \qquad 15.2$$

using 15.1 and 15.2 we get,

$$A_{CL} = \frac{-R_1}{R_2} \qquad 15.3$$

In the present case $R_1 = 10k, R_2 = 5k$

Substituting in 15.3 we get,

$$A_{CL} = \frac{-10k}{5k} = -2$$

Inverting Amplifier using OpAmp

Thus the gain of the inverting amplifier can be adjusted by simply changing the value of resistors R1 and R2. The negative sign just means that the output Vout is 180° out of phase with the input Vin.

For a gain of 2, R1= 10k and R2=5k can be taken as shown in Fig 15.1. Vin is a sinusoidal input of maximum voltage 5V and since the gain is 2, the Vout is a sinusoidal output of maximum voltage 10V. Thus it can be seen that Vout is amplified by a factor of 2 and is 180° out of phase with the inverting input Vin. Va is the Voltage at node a. It can be seen that the voltage level at Va is in the range of a few μVs. Thus Va can be treated as being at ground potential. This is the concept of Virtual ground where the inverting terminal can be treated as being at 0V. This helps in simplifying the circuit analysis and is quite helpful in solving more complicated circuits involving OpAmps.

Inverting Amplifier –the case of Saturation

Fig 15.4 shows the circuit diagram for the inverting amplifier having R1= 1k and R2=10k. Using equation 15.3, the gain comes out to be -10. Vin is a sinusoidal input of maximum voltage 5V and since the gain is -10, the Vout must a sinusoidal output of maximum voltage 50V. But Vout cannot exceed the maximum DC supply which is 15V in the present case. As a result Vout gets clipped and looks more like a square wave as shown in fig 15.5.A proper gain must be chosen for the output signal to remain below the saturation level.

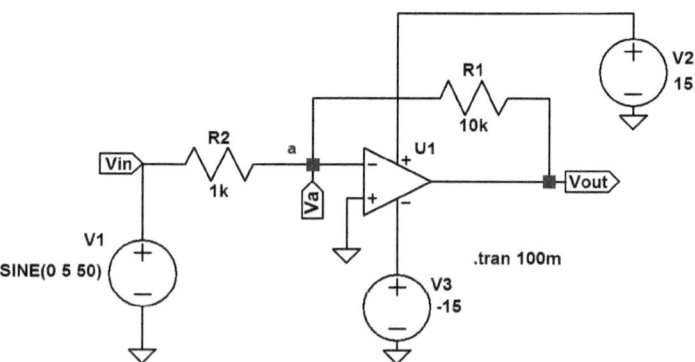

Fig 15.4: Circuit diagram: Inverting Amplifier - the case of Saturation

Electronics Circuit SPICE Simulations with LTspice

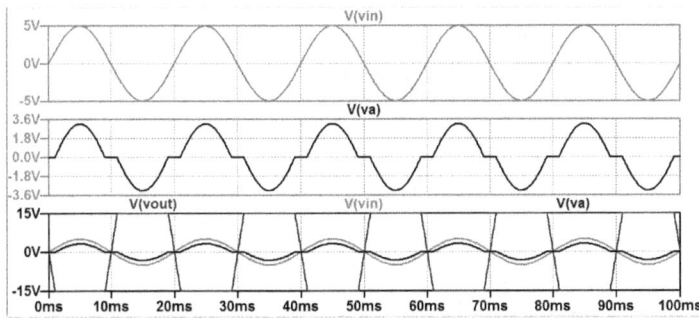

Fig 15.5: Transient Analysis plots (Voltage levels) - Inverting Amplifier

Inverting Amplifier- Ideal model using Voltage Controlled Voltage Controlled voltage Source.

A voltage Controlled Voltage Controlled voltage Source(VCVS) can be used to develop an ideal model of an OpAmp. Fig 15.8 shows an ideal inverting amplifier using a Voltage Controlled Voltage Source (VCVS) model .Voltage Controlled voltage Source (VCVS) can be found in the LTspice component window with a name 'e' as shown in Fig 15.6. VCVS has two inputs and two outputs.

Inverting Amplifier using OpAmp

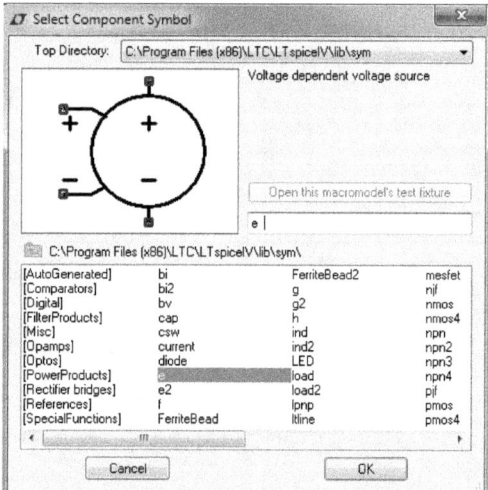

Fig: 15.6 VCVS in component window

Right click on the symbol after placing it on the schematic and enter a gain value for the value attribute, 1e6 in the present case as shown in Fig 15.7. This represents the gain of the OpAmp. As the gain of an OpAmp is dependent on the frequency one may set it to some other lower value, if the gain at that frequency is known. Fig 15.9 shows the Transient Analysis plots for the Inverting Amplifier using a Voltage Controlled Voltage Source (VCVS) model.

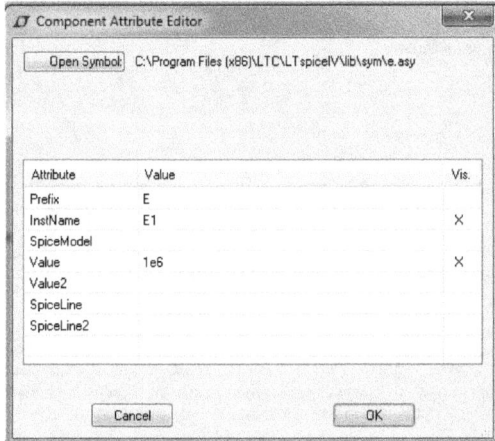

Fig 15.7: Entering a gain value for the value attribute of the VCVS.

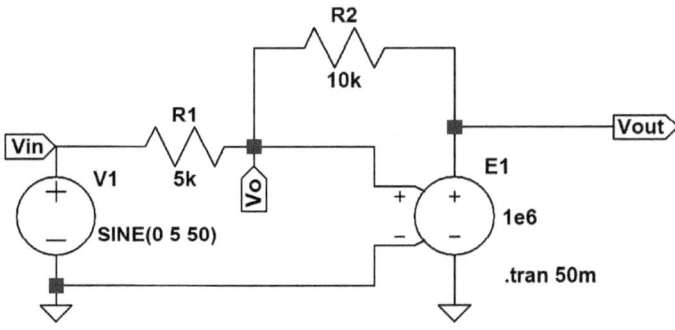

Fig 15.8: An ideal inverting amplifier using a Voltage Controlled Voltage Source (VCVS) model

Inverting Amplifier using OpAmp

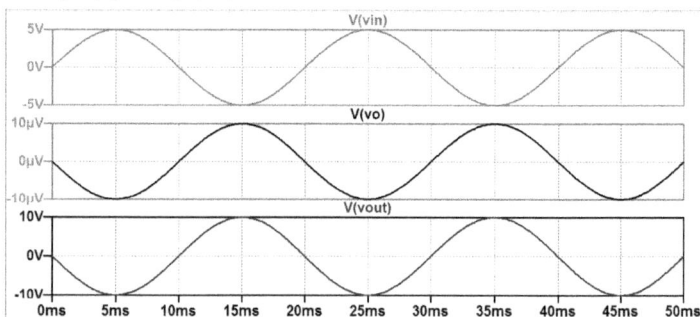

Fig 15.9: Transient Analysis plots (Voltage levels) - Inverting Amplifier using a Voltage Controlled Voltage Source (VCVS) model

2) Non- inverting Amplifier: The input signal is applied only to Non-Inverting terminal and the Inverting terminal is grounded via a resistive path. Non-Inverting Amplifier has been discussed in detail in the next Chapter.

****

Chapter-16   Non-inverting Amplifier using OpAmp

The concept of Non-inverting Amplifier using OpAmp

In a non-inverting amplifier, the input signal is applied to the non-inverting input of the OpAmp. The inverting input of the OpAmp is kept grounded via a resistor. The Output terminal of the OpAmp is connected to the inverting input terminal via a resistor of suitable value. Fig 16.1 shows the circuit diagram for a non-inverting amplifier, R1 is the feedback resistor and R2 is the input resistor. A sinusoidal input of maximum voltage=1V has been applied at the non-inverting input of the amplifier circuit. Fig 16.2 shows the transient analysis plots for various Voltage levels and Fig 16.3 shows Transient Analysis plots for various current levels in the non-inverting amplifier circuit.

The most important property of an amplifier is its Voltage gain. The closed loop voltage gain of the non-inverting amplifier has been calculated in the next section.

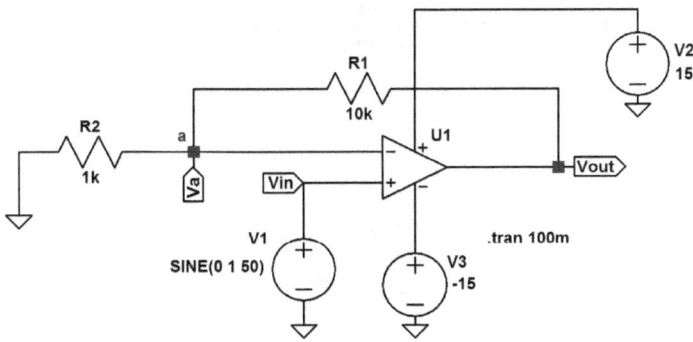

Fig 16.1: Circuit diagram: Non-inverting Amplifier

Non-inverting Amplifier using OpAmp

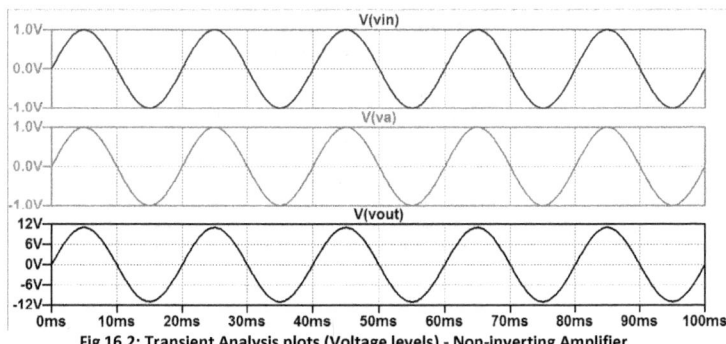

Fig 16.2: Transient Analysis plots (Voltage levels) - Non-inverting Amplifier

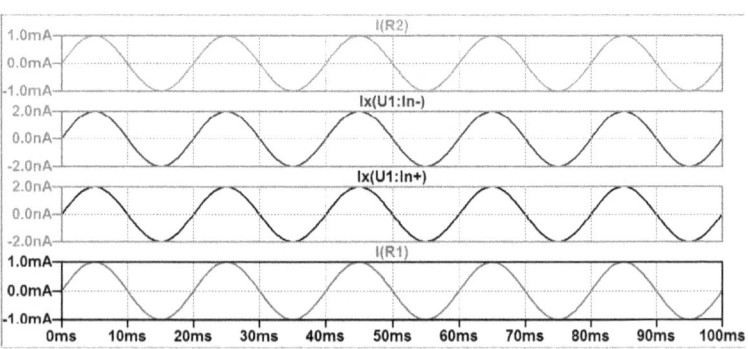

Fig 16.3: Transient Analysis plots (current levels) - Non-inverting Amplifier

Non-inverting amplifier-Closed Loop Voltage Gain

The gain of an non-inverting amplifier can be found out by applying nodal analysis at node a.

Applying Nodal Analysis at Node a, we have,

$$\frac{V_a - 0}{R_2} + \frac{V_a - V_{out}}{R_1} = 0 \qquad\qquad 15.1$$

where V_a represents the voltage at node a.

Now, since V_a is at the same potential as V_{in}.

$$V_a = V_{in}$$

Substituting the value of V_a in 15.1 we get,

$$\frac{V_{in}}{R_2} + \frac{V_{in} - V_{out}}{R_1} = 0$$

$$\Rightarrow \frac{V_{out}}{R_1} = \frac{V_{in}}{R_2} + \frac{V_{in}}{R_1}$$

$$or, \frac{V_{out}}{V_{in}} = 1 + \frac{R_1}{R_2} \qquad\qquad 15.2$$

The closed loop gain A_{CL} of an amplifier is given by :

$$A_{CL} = \frac{V_{out}}{V_{in}} \qquad\qquad 15.3$$

using 15.2 and 15.3 we get,

$$A_{CL} = 1 + \frac{R_1}{R_2} \qquad\qquad 15.4$$

In the present case $R_1 = 10k, R_2 = 1k$

Substituting in 15.3 we get,

$$A_{CL} = 1 + \frac{10k}{1k} = 11$$

Thus the gain of the non-inverting amplifier can be adjusted by simply changing the value of resistors R1 and R2. Unlike the inverting amplifier where the Vout is $180°$ out of phase with the input Vin, in a non-inverting amplifier the Vout remains in the same phase as the input signal. Also the gain is higher by 1 unit as compared to the inverting amplifier for the same values of R1 and R2.

For a gain of 11, R1= 10k and R2=1k can be taken as shown in Fig 16.1. Vin is a sinusoidal input of maximum voltage 1V and since the gain is 11, the Vout is a sinusoidal output of maximum voltage 11V. Thus it can be seen that Vout is amplified by a factor of 11 an is in phase with the non-inverting input Vin. Va is the Voltage at node a. It can be seen that the voltage level at Va is same as that of Vin.

Non-inverting Amplifier –the case of Saturation

Fig 16.4 shows the circuit diagram for the non-inverting amplifier having R1= 1k and R2=10k. Using equation 16.3, the closed loop gain comes out to be 11. Vin is a sinusoidal input of maximum voltage 5V and since the gain is 11, the Vout must a sinusoidal output of maximum voltage 55V. But Vout cannot exceed the maximum DC supply which is 15V in the present case.

Non-inverting Amplifier using OpAmp

As a result Vout gets clipped and looks more like a square wave as shown in fig 16.5. A proper gain must be chosen for the output signal to remain below the saturation level.

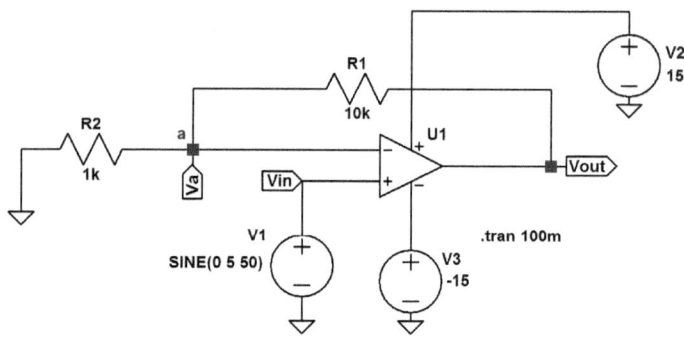

Fig 16.4: Circuit diagram: Non-inverting Amplifier - the case of Saturation

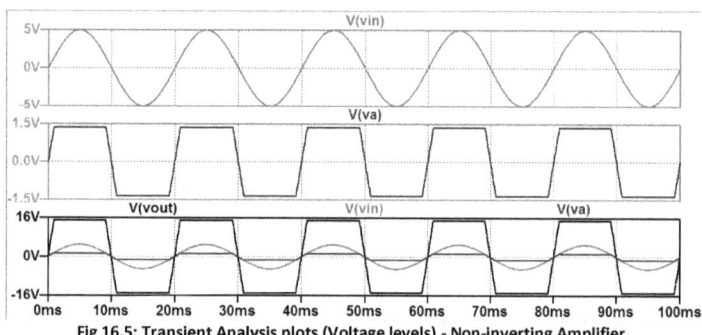

Fig 16.5: Transient Analysis plots (Voltage levels) - Non-inverting Amplifier

Non-inverting Amplifier- Ideal model using Voltage Controlled Voltage Controlled voltage Source.

A voltage Controlled Voltage Controlled voltage Source(VCVS) can be used to develop an ideal model of an OpAmp. Fig 16.6 shows an ideal non-inverting amplifier using a Voltage Controlled Voltage Source (VCVS) model. Fig 16.7 shows the Transient Analysis plots (Voltage levels) for the Non-inverting Amplifier using a Voltage Controlled Voltage Source (VCVS) model.

Electronics Circuit SPICE Simulations with LTspice

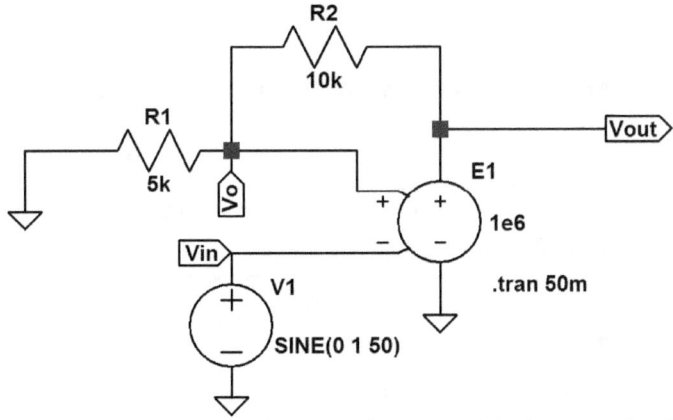

Fig 16.6: An ideal non-inverting amplifier using a Voltage Controlled Voltage Source (VCVS) model

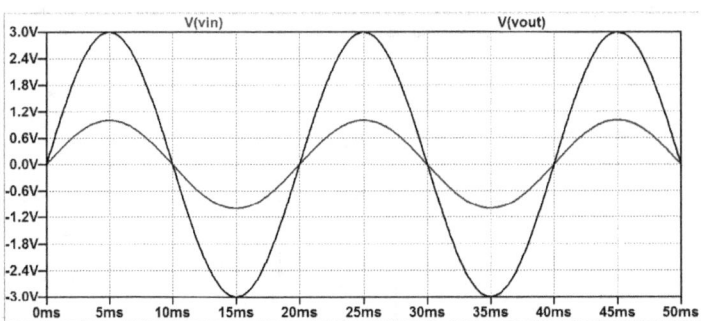

Fig 16.7: Transient Analysis plots (Voltage levels) - Non-inverting Amplifier using a Voltage Controlled Voltage Source (VCVS) model

****

Chapter-17   Voltage summer using OpAmp

Summation operation
In a Summation operation two or more signal are added together. The summation may be of AC signals or DC signals. An OpAmp can work efficiently as a Voltage Summer.

The concept of Voltage Summer using OpAmp
In a Voltage Summer, all the input signals that are to be summed together are applied to the inverting input of the OpAmp. The non-inverting input of the OpAmp is kept grounded. The Output terminal of the OpAmp is connected to the inverting input terminal via a resistor of suitable value.

Voltage Summer- Sine waves as input
Fig 17.1 shows the circuit diagram for a Voltage Summer with R1=2k, R2=1k R3=2k R4=3k where R2, R3 and R4 are the input resistors and R1 is the feedback resistor. Sinusoidal inputs of maximum voltage vin1=4V, vin2=3V and vin3= 2V have been applied at the inverting input of the OpAmp. Fig 17.2 shows the transient analysis plots for various Voltage levels of the Summer circuit in Fig 17.1. Fig 17.3 shows the circuit diagram for a Summer circuit with R1=R2=R3=R4=5k. Fig 17.4 shows the transient analysis plots for various Voltage levels for the circuit shown in Fig 17.3.

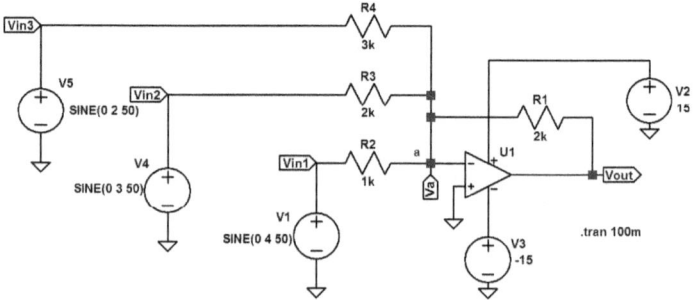

Fig 17.1: Circuit diagram: Voltage Summer with R1=2k, R2= 1k, R3=2k, R4=3k (Sinusoidal Inputs)

Voltage summer using OpAmp

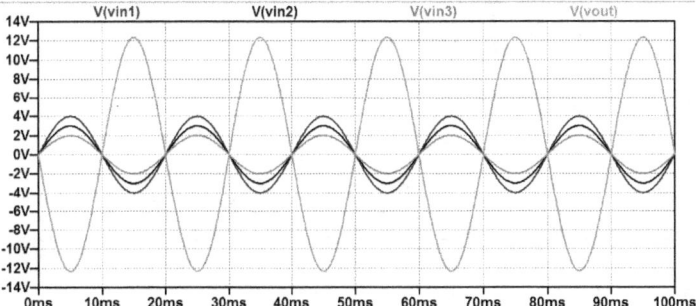

Fig 17.2: Transient Analysis plots (Voltage levels) – Voltage Summer

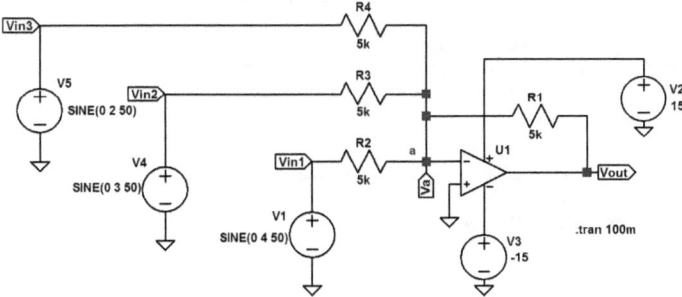

Fig 17.3: Circuit diagram: Voltage Summer with R1=R2=R3=R4=5k

Electronics Circuit SPICE Simulations with LTspice

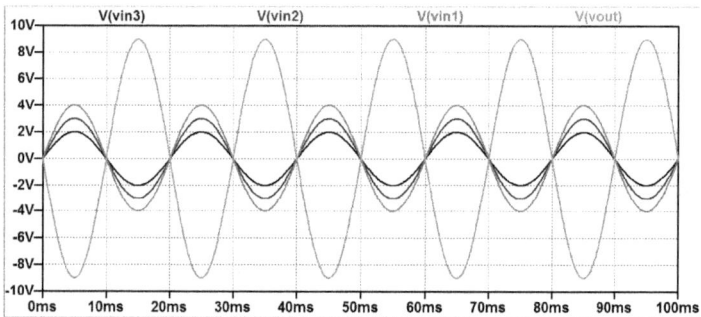

Fig 17.4: Transient Analysis plots (Voltage levels) – Voltage Summer

Voltage Summer- Output Voltage

The output Voltage of a Summer can be found out by applying nodal analysis at node a.

Applying Nodal Analysis at Node a, we have,

$$\frac{V_a - V_{in1}}{R_2} + \frac{V_a - V_{in2}}{R_3} + \frac{V_a - V_{in3}}{R_4} + \frac{V_a - V_{out}}{R_1} = 0 \qquad 17.1$$

where V_a represents the voltage at node a.

Now, since V_a is at the same potential as the ground(Virtual ground).

$$V_a = 0$$

Substituting the value of V_a in 17.1 we get,

$$\frac{-V_{in1}}{R_2} + \frac{-V_{in2}}{R_3} + \frac{-V_{in3}}{R_4} + \frac{-V_{out}}{R_1} = 0$$

$$\Rightarrow \frac{V_{out}}{R_1} = -\left(\frac{V_{in1}}{R_2} + \frac{V_{in2}}{R_3} + \frac{V_{in3}}{R_4}\right)$$

$$\text{or, } V_{out} = -R_1\left(\frac{V_{in1}}{R_2} + \frac{V_{in2}}{R_3} + \frac{V_{in3}}{R_4}\right)$$

$$\text{or, } V_{out} = -\left(\frac{R_1}{R_2}V_{in1} + \frac{R_1}{R_3}V_{in2} + \frac{R_1}{R_4}V_{in3}\right) \qquad 17.2$$

Thus V_{out} is the sum of input Voltages when we choose resistances such that,

$$R_1 = R_2 = R_3 = R_4.$$

In the present case $R_1 = 2k, R_2 = 1k, R_3 = 2k, R_4 = 3k$

V_{in1} max = 4V, V_{in2} max = 3V, V_{in3} max = 2V

Substituting in 17.2 we get,

$$V_{out}\text{ max} = -\left(\frac{2k}{1k}4 + \frac{2k}{2k}3 + \frac{2k}{3k}2\right)$$

$$V_{out}\text{ max} = -(8 + 3 + 1.33)$$

$$V_{out}\text{ max} = -12.33V$$

Voltage summer using OpAmp

Thus the Summation output(max) for the circuit in Fig 17.1 comes out to be -12.33 . The negative sign just means that the output Vout is $180°$ out of phase with the inputs.

Voltage Summer- DC inputs

Fig 17.5 shows the circuit diagram for a Summer circuit with R1=R2=R3=R4=5k with DC inputs vin1=4V, vin2=3V and vin3= 2V. Fig 17.6 shows the transient analysis plots for various Voltage levels for the circuit shown in Fig 17.5. Using equation 17.2, the output comes out to be −9V.

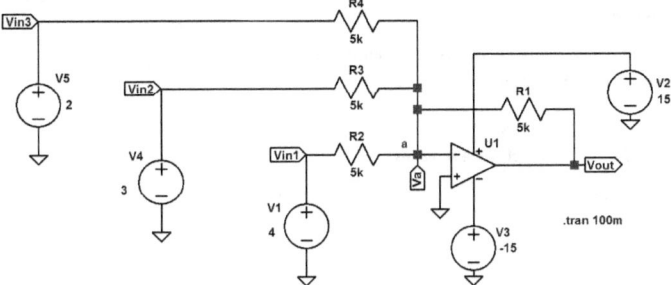

Fig 17.5: Circuit diagram: Voltage Summer with R1=R2=R3=R4=5k (DC Inputs)

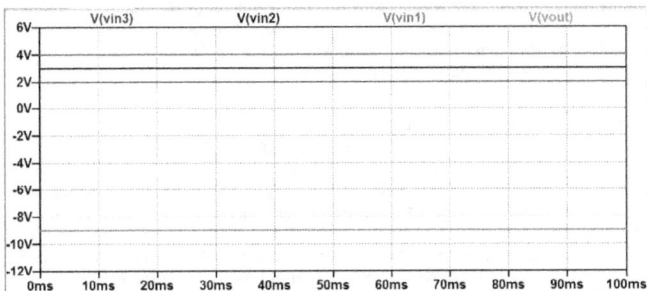

Fig 17.6: Transient Analysis plots (Voltage levels) – Voltage Summer(DC Inputs)

****

Chapter-18   Differentiator using OpAmp

Differentiation operation

It is known that differentiation of a function is its rate of change with respect to time. In many applications differentiation operation is desired. An OpAmp can work efficiently as a differentiator.

Lets look at a few simple examples of differentiation.

a) Equation of a line is given as: $y = ax + c$

$$\frac{dy}{dx} = a$$

thus, the differentiation of a line is a constant.

b) $\dfrac{d(\sin x)}{dx} = \cos x$

c) $\dfrac{d(\cos x)}{dx} = -\sin x$

The concept of Differentiator using OpAmp

In a Differentiator, the input signal is applied to the inverting input of the OpAmp. The non-inverting input of the OpAmp is kept grounded. The Output terminal of the OpAmp is connected to the inverting input terminal via a resistor of suitable value while the input capacitor of suitable value is connected at the inverting input.

Differentiator- Sine wave as input

Fig 18.1 shows the circuit diagram for a Differentiator with R1=1k, C1=0.16μ where R1 is the feedback resistor and C1 is the input Capacitor. A sinusoidal input of maximum voltage=1V has been applied at the inverting input of the amplifier circuit. Fig 18.2 shows the transient analysis plots for various Voltage levels and Fig 18.3 shows Transient Analysis plots for various current levels for the Differentiator circuit in Fig 18.1.

Fig 18.4 shows the circuit diagram for a Differentiator circuit with R1=1K, C1=1.6n. Fig 18.5 shows the transient analysis plots for various Voltage levels for the circuit shown in Fig 18.4.

Differentiator using OpAmp

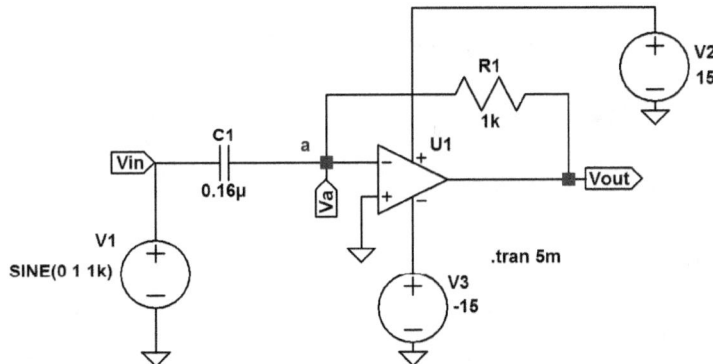

Fig 18.1: Circuit diagram: Differentiator- Sine wave as input(R1=1k, C1=0.16μ)

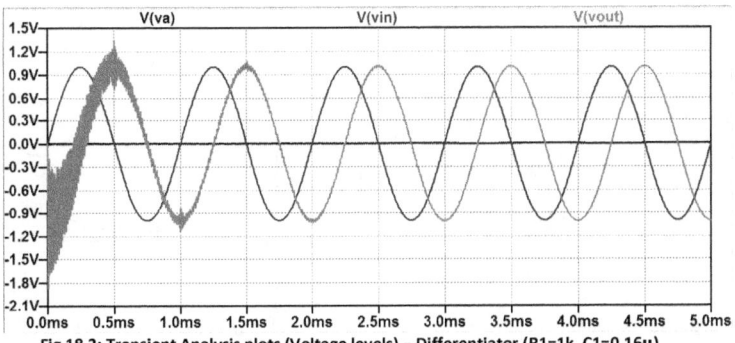

Fig 18.2: Transient Analysis plots (Voltage levels) – Differentiator (R1=1k, C1=0.16μ)

Electronics Circuit SPICE Simulations with LTspice

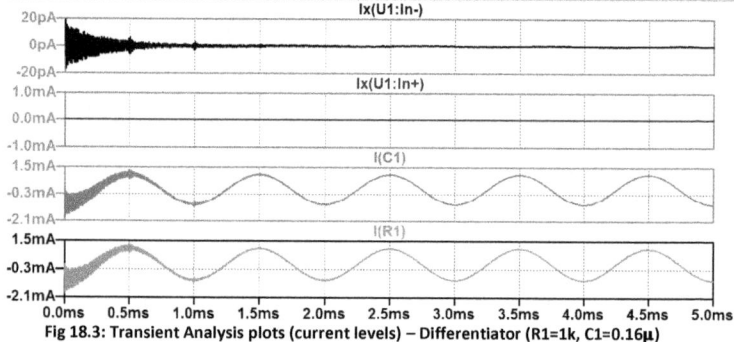

Fig 18.3: Transient Analysis plots (current levels) – Differentiator (R1=1k, C1=0.16µ)

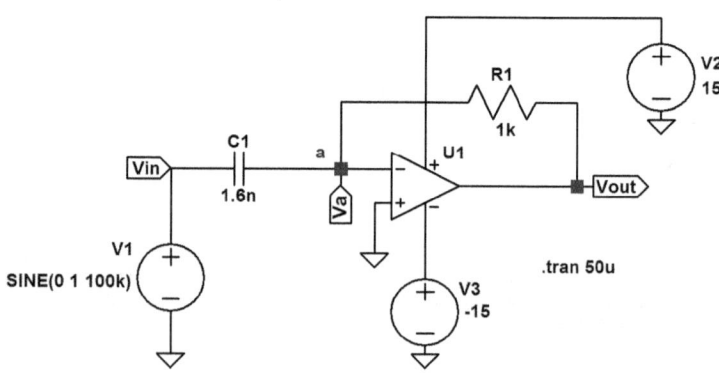

Fig 18.4: Circuit diagram: Differentiator- Sine wave as input (R1=1k, C1=1.6n)

Differentiator using OpAmp

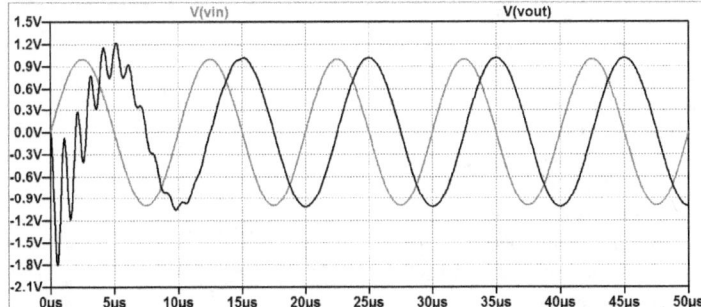

Fig 18.5: Transient Analysis plots (Voltage levels) – Differentiator (R1=1k, C1=1.6n)

Differentiator- Output Voltage

The output Voltage of a Differentiator can be found out by using the Voltage-current relationship of a capacitor.

The current flowing through a capacitor I_c is given by,

$$I_c = C \frac{dV_c}{dt} \qquad\qquad 18.1$$

where V_c = Voltage across the capacitor.

Applying Nodal Analysis at Node a, we have,

$$\frac{V_a - V_{out}}{R_1} + C_1 \frac{d(V_a - V_{in})}{dt} = 0 \qquad\qquad 18.2$$

where V_a represents the voltage at node a.

Now, since V_a is at the same potential as the ground(Virtual ground).

$$V_a = 0$$

Substituting the value of V_a in 18.2 we get,

$$\frac{-V_{out}}{R_1} + C_1 \frac{d(-V_{in})}{dt} = 0$$

$$or, \frac{-V_{out}}{R_1} - C_1 \frac{d(V_{in})}{dt} = 0$$

$$\Rightarrow V_{out} = -R_1 C_1 \frac{d(V_{in})}{dt} \qquad\qquad 18.3$$

Thus, V_{out} is the differentiation of V_{in} with a scaling factor of $-R_1 C_1$.

Electronics Circuit SPICE Simulations with LTspice

In the present case $R_1 = 1k$, $C_1 = 0.16\mu$

$V_{in} = \sin(2\pi \cdot 1000 \cdot t)$

Substituting in 18.3 we get

$V_{out} = -1.6 \times 10^{-4} \left(\dfrac{d(\sin(2\pi \cdot 1000 \cdot t))}{dt} \right)$

$V_{out} = -1.6 \times 10^{-4} \left(\cos(2\pi \cdot 1000 \cdot t) \right) \left(2\pi \cdot 1000 \cdot t \right)$

$V_{out} = -1.6 \times 10^{-4} \left(\cos(2\pi \cdot 1000 \cdot t) \right) \left(2\pi \cdot 1000 \right)$

$V_{out} = -10.048 \times 10^{3} \times 10^{-4} \left(\cos(2\pi \cdot 1000 \cdot t) \right)$

$V_{out} = -\left(\cos(2\pi \cdot 1000 \cdot t) \right)$ 18.4

Thus the output of the Differentiator can be adjusted by changing the value of resistor R1 and capacitor C1. The negative sign just means that the output Vout is $180°$ out of phase with the differentiation of Vin.

Differentiator – triangular wave as input

Fig 18.6 shows the circuit diagram for a Differentiator having R1= 5k and C1=10n. Vin is a triangular input of maximum voltage 1V and thus Vout is a square wave output of maximum voltage 50mV.

In the present case $R_1 = 5k$, $C_1 = 10n$

$V_{in} = \begin{cases} 1000t & \text{for } 0 \leq t < 1ms \\ -1000t & \text{for } 1ms \leq t < 2ms \end{cases}$ 18.5

Substituting in 18.3 we get

$V_{out} = -5 \times 10^{3} \times 10^{-8} \left(\dfrac{d(1000t)}{dt} \right)$ for $0 \leq t < 1ms$

$V_{out} = -5 \times 10^{3} \times 10^{-8} \times 1000$

$V_{out} = -5 \times 10^{-2}$

or, $V_{out} = -50\ mV$ for $0 \leq t < 1ms$

similarly,

$V_{out} = 50\ mV$ for $1ms \leq t < 2ms$ 18.6

Fig 18.7 shows the Transient Analysis plots (Voltage levels) for the Differentiator of Fig 18.6. Fig 18.8 shows the Circuit diagram for a Differentiator having R1= 5k and C1=5n. Fig 18.9 shows the Transient Analysis plots (Voltage levels) for the Differentiator of Fig 18.8.

Differentiator using OpAmp

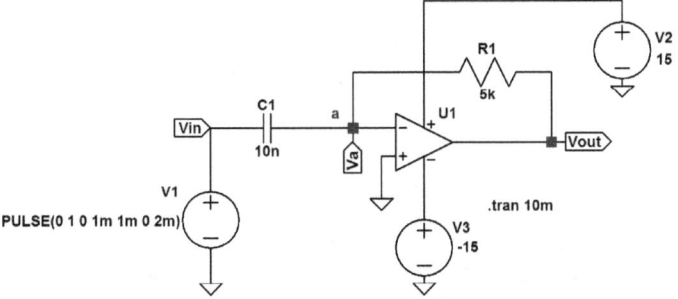

Fig 18.6: Circuit diagram: Differentiator – triangular wave as input (R1= 5k and C1=10n)

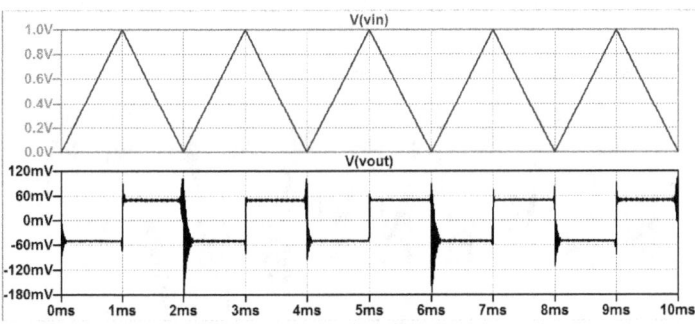

Fig 18.7: Transient Analysis plots (Voltage levels) – Differentiator- triangular wave as input (R1= 5k and C1=10n)

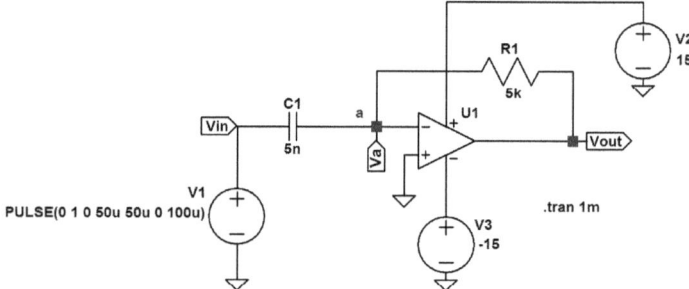

Fig 18.8: Circuit diagram: Differentiator – triangular wave as input (R1= 5k and C1=5n)

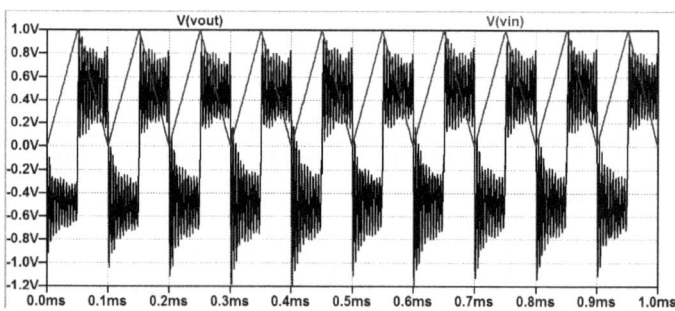

Fig 18.9: Transient Analysis plots (Voltage levels) – Differentiator- triangular wave as input
(R1= 5k and C1=5n)

Differentiator- AC Analysis

Fig 18.10 shows a Differentiator circuit having R1= 1k, C1=0.16μ with AC Analysis command in the frequency range 1 kHz to 130 kHz. AC Analysis shows that the differentiator circuit can work like an Active high pass filter as it suppresses the lower frequencies and allows the higher frequencies to pass. Although the High pass response is not ideal but it works for quite a large range of high frequencies. The range of frequencies that are to be allowed to pass can be altered by properly choosing the value of R1 and C1. Fig 18.11 shows the AC Analysis plots(Magnitude and phase) for the circuit in Fig 18.10. Fig 18.12 shows a Differentiator circuit

Differentiator using OpAmp

having R1= 5k, C1=5n with AC Analysis command in the frequency range 100 kHz to 1MegaHz.
Fig 18.13 shows the AC Analysis plots (magnitude and phase) for the circuit in Fig 18.12.

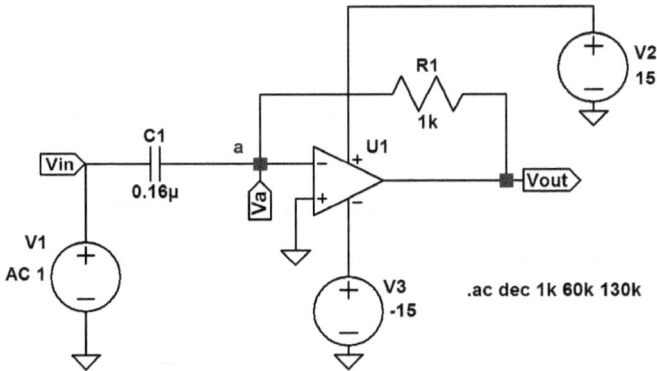

Fig 18.10: Circuit diagram: Differentiator (R1= 1k, C1=0.16μ) – AC Analysis

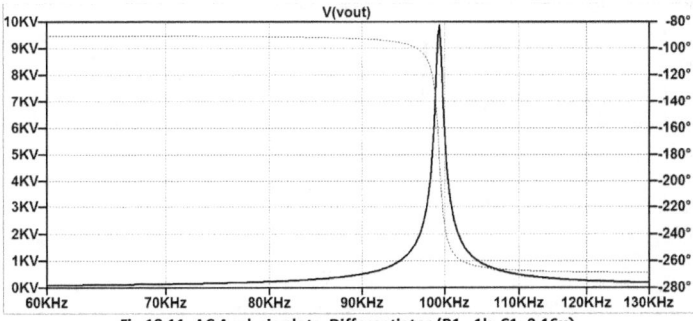

Fig 18.11: AC Analysis plots: Differentiator (R1= 1k, C1=0.16μ)

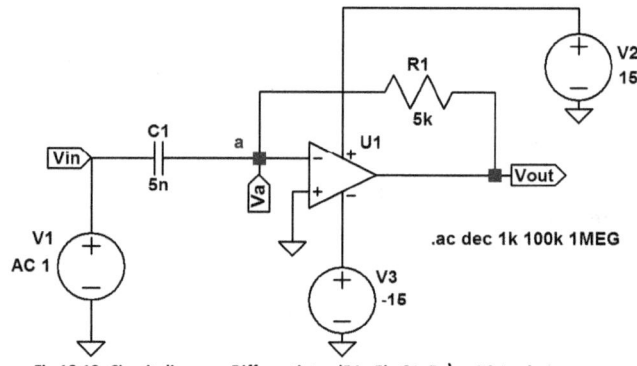

Fig 18.12: Circuit diagram: Differentiator (R1= 5k, C1=5n) − AC Analysis

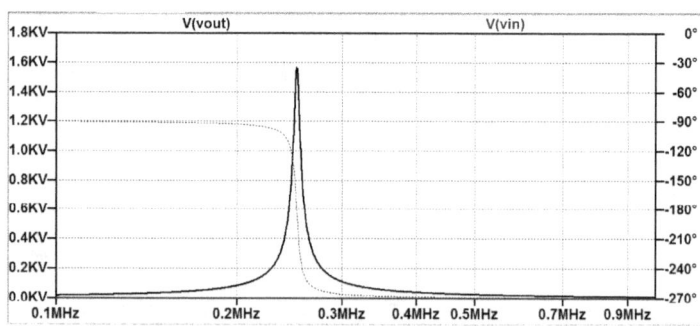

Fig 18.13: AC Analysis plots: Differentiator (R1= 5k, C1=5n)

****

Chapter-19   Integrator using OpAmp

Integration operation

If the rate of change of a function with respect to time is known, Integration gives back the original function. In many applications Integration operation is desired. An OpAmp can work efficiently as an Integrator.

Lets look at a few simple examples of Integration.

a) Equation of a line is given as : $y = ax + c$

$$\int y\,dx = a\frac{x^2}{2} + cx$$

b) $\int \sin x = -\cos x$

c) $\int \cos x = \sin x$

The concept of Integrator using OpAmp

In an Integrator, the input signal is applied to the inverting input of the OpAmp. The non-inverting input of the OpAmp is kept grounded. The Output terminal of the OpAmp is connected to the inverting input terminal via a capacitor of suitable value while the input resistor of suitable value is connected at the inverting input.

Integrator- Sine wave as input

Fig 19.1 shows the circuit diagram for an Integrator with R1=20k, C1=10n where R1 is the input resistor and C1 is the feedback Capacitor. A sinusoidal input of maximum voltage=1V has been applied at the inverting input of the amplifier circuit. Fig 19.2 shows the transient analysis plots for various Voltage levels and Fig 19.3 shows Transient Analysis plots for various current levels for the Integrator circuit in Fig 19.1.

Fig 19.4 shows the circuit diagram for an Integrator circuit with R1=200k, C1=5n. Fig 19.5 shows the transient analysis plots for various Voltage levels for the circuit shown in Fig 19.4.

Electronics Circuit SPICE Simulations with LTspice

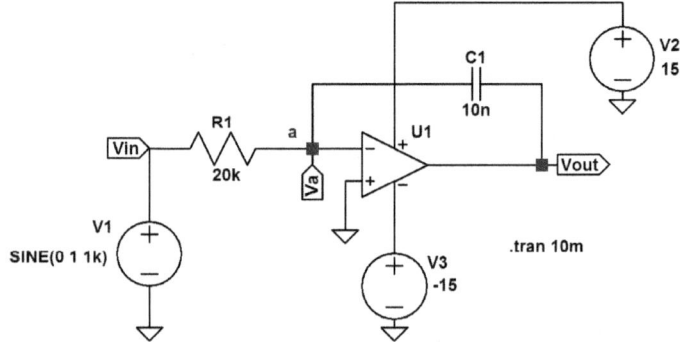

Fig 19.1: Circuit diagram: Integrator- Sine wave as input (R1=20k, C1=10n)

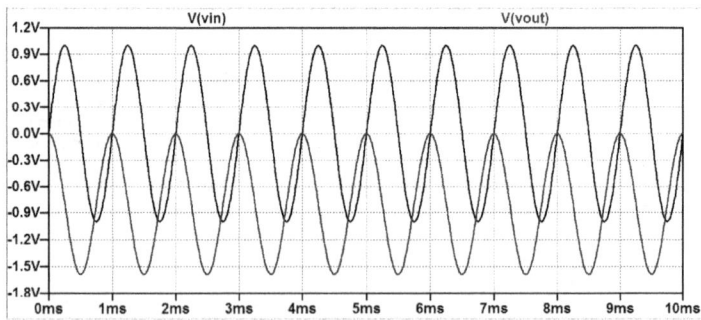

Fig 19.2: Transient Analysis plots (Voltage levels) – Integrator (R1=20k, C1=10n)

124

Integrator using OpAmp

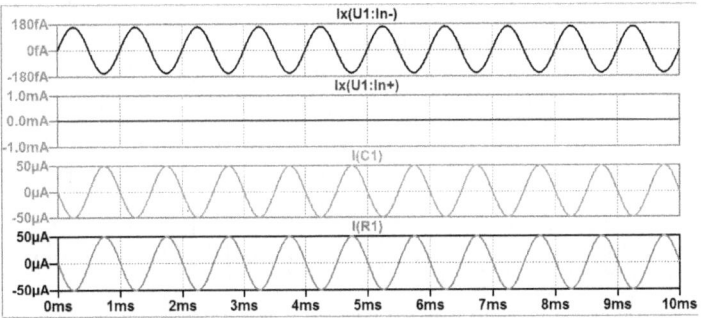

Fig 19.3: Transient Analysis plots (current levels) – Integrator (R1=20k, C1=10n)

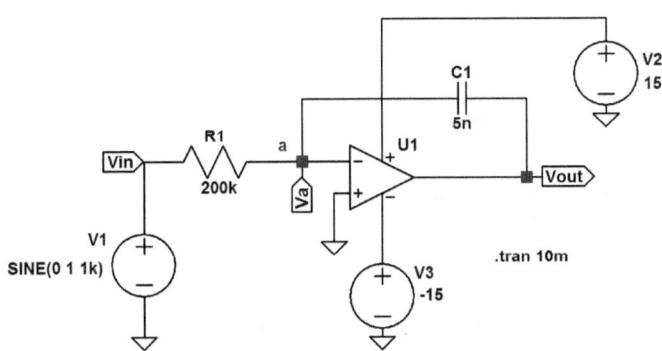

Fig 19.4: Circuit diagram: Integrator- Sine wave as input (R1=200k, C1=5n)

Electronics Circuit SPICE Simulations with LTspice

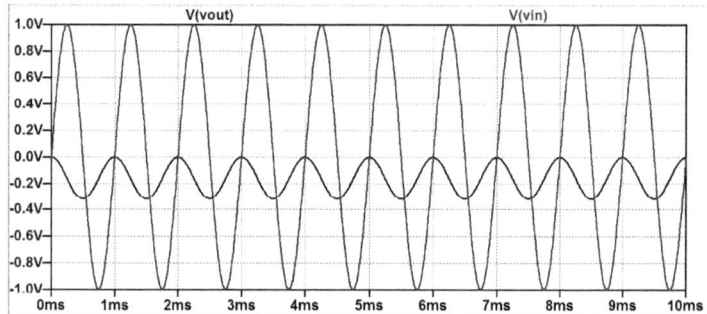

Fig 19.5: Transient Analysis plots (Voltage levels) – Integrator (R1=200k, C1=5n)

Integrator- Output Voltage

The output Voltage of a Integrator can be found out by using the Voltage-current relationship of a capacitor.

The current flowing through a capacitor I_c is given by,

$$I_c = C \frac{dV_c}{dt}$$ 18.1

where V_c = Voltage across the capacitor.

Applying Nodal Analysis at Node a, we have,

$$\frac{V_a - V_{in}}{R_1} + C_1 \frac{d(V_a - V_{out})}{dt} = 0$$ 18.2

where V_a represents the voltage at node a.

Now, since V_a is at the same potential as the ground(Virtual ground).

$$V_a = 0$$

Substituting the value of V_a in 18.2 we get,

$$\frac{-V_{in}}{R_1} + C_1 \frac{d(-V_{out})}{dt} = 0$$

$$or, \frac{-V_{in}}{R_1} - C_1 \frac{d(V_{out})}{dt} = 0$$

Integrator using OpAmp

or, $-C_1 \dfrac{d(V_{out})}{dt} = \dfrac{V_{in}}{R_1}$

or, $\dfrac{d(V_{out})}{dt} = \dfrac{V_{in}}{-R_1 C_1}$

or, $d(V_{out}) = \dfrac{V_{in}}{-R_1 C_1} dt$ 18.3

Integarting on both sides of 18.3 we get,

$$\Rightarrow V_{out} = \dfrac{1}{-R_1 C_1} \int V_{in} dt \qquad\qquad 18.4$$

Thus, V_{out} is the integration of V_{in} with a scaling factor of $-\dfrac{1}{R_1 C_1}$.

In the present case $R_1 = 20k$, $C_1 = 10n$

$V_{in} = \sin(2\pi \bullet 1000 \bullet t)$

Substituting in 18.3 we get

$V_{out} = \dfrac{1}{-2 \times 10^{-4}} \int \sin(2\pi \bullet 1000 \bullet t) dt$

$V_{out} = -5 \times 10^3 \int \sin(2\pi \bullet 1000 \bullet t) dt$

$V_{out} = \dfrac{5 \times 10^3}{2\pi \bullet 1000} (\cos(2\pi \bullet 1000 \bullet t))$

$V_{out} = \dfrac{5}{6.28} (\cos(2\pi \bullet 1000 \bullet t))$

$V_{out} = 0.796 (\cos(2\pi \bullet 1000 \bullet t))$

$V_{out} \approx 0.8 (\cos(2\pi \bullet 1000 \bullet t))$ 18.5

Thus the output of the Integrator can be adjusted by changing the value of resistor R1 and capacitor C1. The positive sign means that the output Vout is in phase with the Integration of Vin.

Integrator – Square wave as input

Fig 19.6 shows the circuit diagram for an Integrator having R1= 100k and C1=10n. Vin is a square wave input of maximum voltage 1V and thus Vout is a triangular wave output of slopes 10^3 and 10^{-3} .

Electronics Circuit SPICE Simulations with LTspice

In the present case $R_1 = 100k$, $C_1 = 10n$

$$V_{in} = \begin{cases} 1 & \text{for } 0 \leq t < 1ms \\ -1 & \text{for } 1ms \leq t < 2ms \end{cases} \qquad 18.6$$

Substituting in 18.4 we get

$$V_{out} = -\frac{1}{10^{-3}} \int 1 \, dt \qquad \text{for } 0 \leq t < 1ms$$

$$V_{out} = -10^3 t \qquad \text{for } 0 \leq t < 1ms$$

similarly,

$$V_{out} = 10^3 t \qquad \text{for } 1ms \leq t < 2ms \qquad 18.7$$

Fig 19.7 shows the Transient Analysis plots (Voltage levels) for the Integrator of Fig 19.6. Fig 19.8 shows the Circuit diagram for a Integrator having R1= 1000k and C1=1n. Fig 19.9 shows the Transient Analysis plots (Voltage levels) for the Integrator of Fig 19.8.

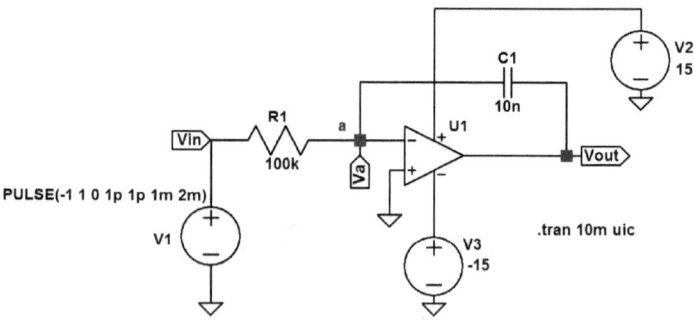

Fig 19.6: Circuit diagram: Integrator – square wave as input (R1= 100k and C1=10n)

Integrator using OpAmp

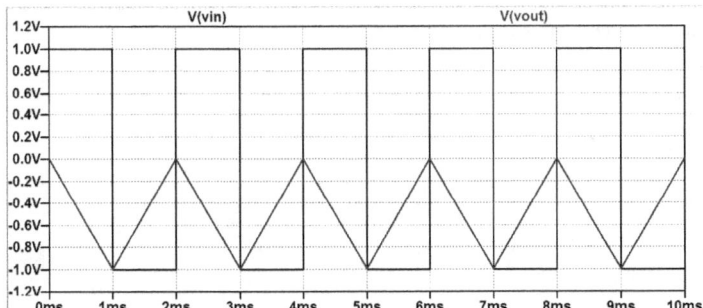

Fig 19.7: Transient Analysis plots (Voltage levels) – Integrator- Square wave as input (R1= 100k and C1=10n)

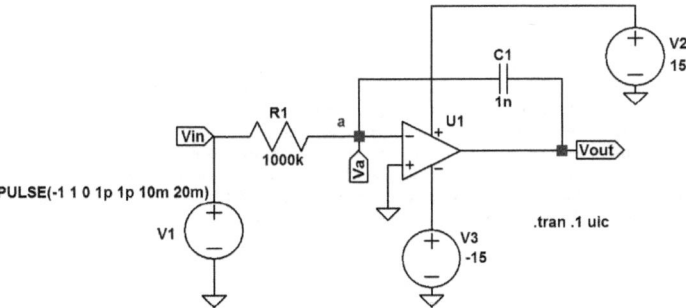

Fig 19.8: Circuit diagram: Integrator – square wave as input (R1= 1000k and C1=1n)

Electronics Circuit SPICE Simulations with LTspice

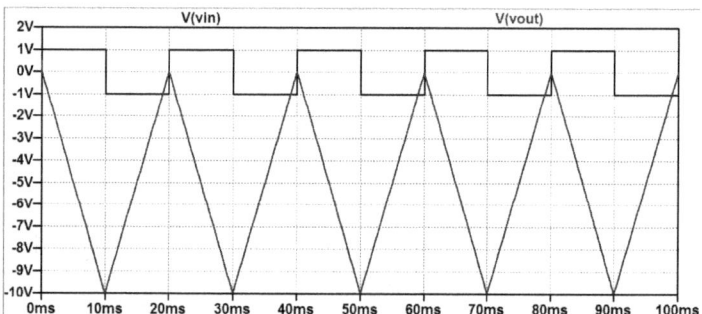

Fig 19.9: Transient Analysis plots (Voltage levels) – Integrator- square wave as input (R1= 1000k and C1=1n)

Integrator- AC Analysis

Fig 19.10 shows an Integrator circuit having R1= 1MEG, C1=1n with AC Analysis command in the frequency range 1 Hz to 1 kHz. AC Analysis shows that the Integrator circuit can work like an Active low pass filter as it suppresses the higher frequencies and allows the lower frequencies to pass. The range of frequencies that are to be allowed to pass can be altered by properly choosing the value of R1 and C1. Fig 19.11 shows the AC Analysis plots(Magnitude and phase) for the circuit in Fig 19.10. Fig 19.12 shows a Integrator circuit having R1= 20k, C1 = 10n with AC Analysis command in the frequency range 1Hz to 1kHz. Fig 19.13 shows the AC Analysis plots (magnitude and phase) for the circuit in Fig 19.12.

Integrator using OpAmp

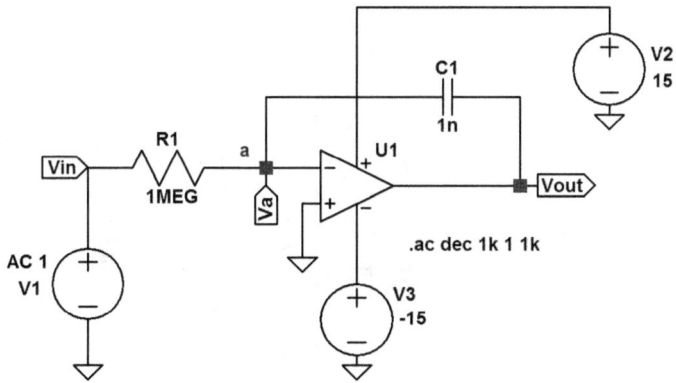

Fig 19.10: Circuit diagram: Integrator (R1= 1MEG, C1=1n) – AC Analysis

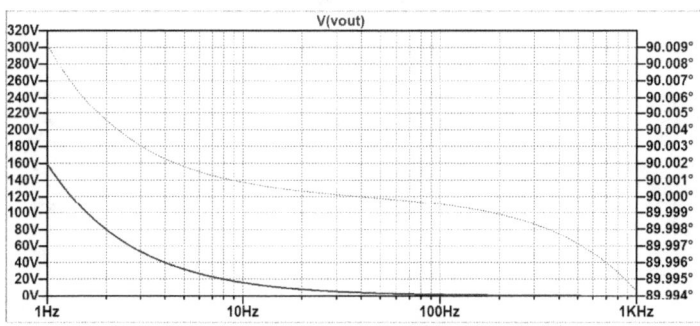

Fig 19.11: AC Analysis plots: Integrator (R1= 1MEG, C1=1n)

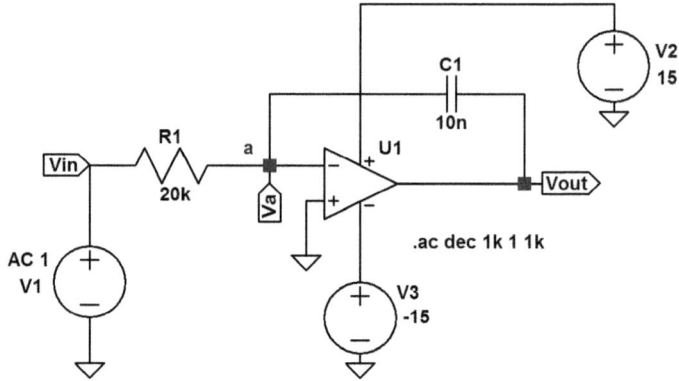

Fig 19.12: Circuit diagram: Integrator (R1= 20k, C1=10n) – AC Analysis

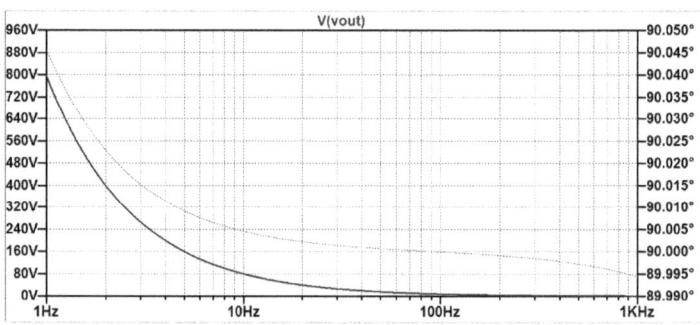

Fig 19.13: AC Analysis plots: Integrator (R1= 20k, C1=10n)

****

Chapter-20   Precision Half-Wave Rectifier using Operational Amplifier

The concept of a Precision Half-Wave Rectifier

A Precision rectifier is like an ideal rectifier consisting of a diode and an operational amplifier. A simple rectifier consisting of a diode doesn't do a perfect rectification. This is because a diode conducts only when the voltage applied across it is greater than the threshold voltage that is 0.7 Volts for a silicon Diode. When the input voltage is smaller than the threshold voltage there in no current across the diode as a result 0V output is obtained. This problem is taken care in the precision rectifier where even for lower voltages ideal rectification is obtained. The opamp due to its high open loop gain(A_{OL}) decreases the threshold voltage of the diode connected in its feedback path by a factor that is equal to the open loop gain of the opamp. As a result highly accurate rectification can be obtained. Precision rectifier finds application in high accuracy signal processing. For small signal circuits where the accuracy required is in the range of fractions of a volt, Precision rectifier is particularly suitable. Only at very low voltages in the range of a few milliVolts the precision rectifier doesn't work well as shown via simulations in Fig 20.3 and Fig 20.6.

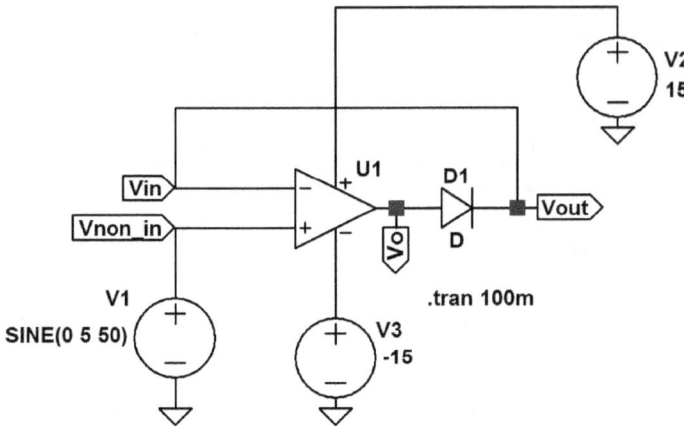

Fig 20.1: A Typical Precision Half-Wave Rectifier with sinusoidal input of amplitude 5V

Precision Half-Wave Rectifier using Operational Amplifier

The Precision Half-Wave Rectifier-Operation

Case-1: Input- Positive half cycle of the Sinusoidal Input

Let us assume that the input voltage is in positive half cycle. A diode is ON only when a voltage greater than the threshold voltage is applied across its terminals. In the positive half of the input cycle when the input voltage is very low, the diode is in the OFF state and the feedback loop is like an open circuit. And the opamp is now working in open loop configuration. It is known that in open loop configuration the opamp has a very high voltage gain. The threshold voltage of the diode gets divided by the open loop gain (A_{OL}) of the operational amplifier. As a result the input voltage is amplified several times and even though the input voltage is low the diode receives a very high input voltage to switch it ON. So, a high voltage appears at Vo as the open loop gain of an opamp is very high. The diode now turns ON at a voltage much lower than 0.7 Volts. Thus even when a very little voltage appears at Vnon_in, a high positive voltage at the anode of the Diode D1 turns it ON. The diode can now be treated like a closed switch. Once the diode is in ON state the circuit works just like a voltage follower circuit. Thus the entire positive half cycle is now available at Vout.

As discussed earlier, an operational amplifier in closed loop configuration tries to keep both the inputs at the same voltage level. For example if there is a change in the voltage level at non-inverting input, the operational amplifier tries to compensate this via changing the current in the feedback path such that the inverting input also arrives at the same voltage level as the present voltage of the non-inverting input . Again this is the principle upon which a voltage follower circuit works. In case of a precision rectifier, when a sinusoidal voltage is applied to the Vnon_in(noninverting input) of the opamp as shown in Fig 20.1, for the positive half cycle of the sinusoidal input when the diode is forward biased it conducts. As a result the diode can be replaced by a short circuit and the feedback loop can simply be seen as that of the Voltage follower circuit.

Case-2: Input- Negative half cycle of the Sinusoidal Input

When the input voltage is in the negative half cycle of the sinusoidal input, Vo becomes negative. A high negative voltage at the anode of the Diode D1 turns it off as the diode is now reverse biased. The diode D1 behaves like an open switch. The opamp is now working in open loop configuration. Vnon_in is amplified by the open loop gain (A_{OL}) of the operational amplifier and Vo hits its maximum negative value that is equal to nearly -15V in the present case. Since the feedback path is now open, no current can flow and thus Vout = 0V. Fig 20.2 shows the Transient Analysis plots for the Precision Half-Wave Rectifier with a sinusoidal input of amplitude 5V.

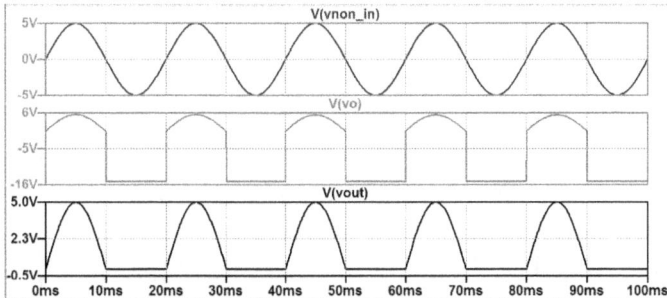

Fig 20.2: Rectified output-Transient Analysis plots-Precision Half-Wave Rectifier with sinusoidal input of amplitude 5V

The Precision Half-Wave Rectifier- Small Signal Input of 20mV

Fig 20.3 shows the Precision Half-Wave Rectifier with sinusoidal input of amplitude 20mV. Fig 20.4 shows the transient analysis plots with a comparison of Vo and Vnon_in in the positive half cycle. Fig 20.5 shows the transient analysis plots vnon_in, Vo and Vout. It can be seen that for such a low voltage the precision rectifier doesn't rectify perfectly.

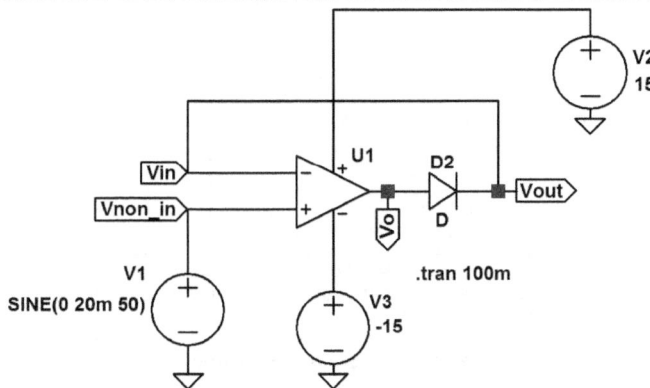

Fig 20.3: A Typical Precision Half-Wave Rectifier with sinusoidal input of amplitude 20mV

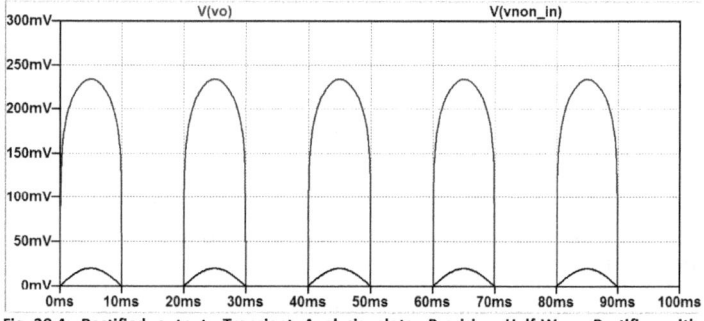

Fig 20.4: Rectified output- Transient Analysis plots- Precision Half-Wave Rectifier with sinusoidal input of amplitude 20mV (comparison of Vo and Vnon_in in the positive half cycle)

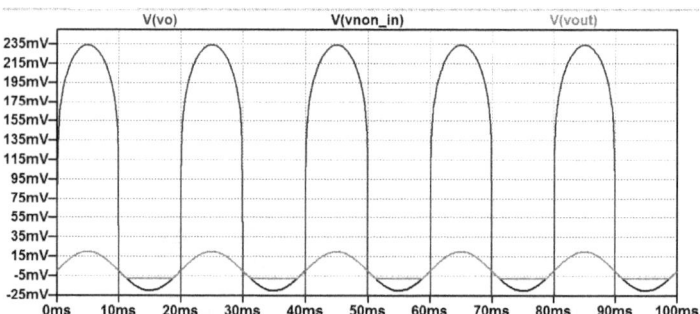

Fig 20.5: Rectified output- Transient Analysis plots- Precision Half-Wave Rectifier with sinusoidal input of amplitude 20mV

The Precision Half-Wave Rectifier- Small Signal Input of 10mV

Fig 20.6 shows the Precision Half-Wave Rectifier with sinusoidal input of amplitude 10mV. Fig 20.7 shows the transient Analysis plots for the Precision Half-Wave Rectifier with a sinusoidal input. It can be seen that as Vnon_in is lowered further the precision rectifier performs even poorly as shown via simulations in Fig 20.7.

Electronics Circuit SPICE Simulations with LTspice

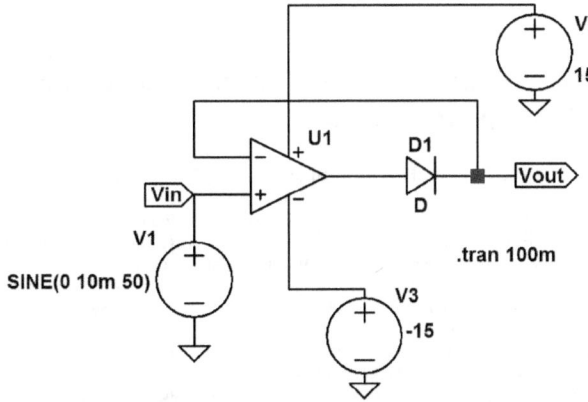

Fig 20.6: A Typical Precision Half-Wave Rectifier with sinusoidal input of amplitude 10mV

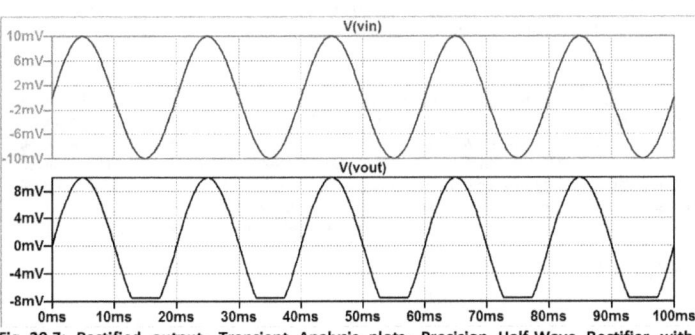

Fig 20.7: Rectified output- Transient Analysis plots- Precision Half-Wave Rectifier with sinusoidal input of amplitude 10mV

****

Chapter 21   Capacitor Charging and Discharging (RC Circuit - Time Response)

The concept of a RC Circuit (The case of DC Supply Voltage)

Capacitors are used to store electrical energy. When a DC Voltage is applied to a Capacitor, it charges to the value of voltage that is applied across its terminals, i.e. if 5V is applied across a Capacitor, the capacitor charges to 5V almost instantly. However, almost always the capacitor is connected to the Voltage supply via a resistive path. As a result when a voltage is applied to the circuit as shown in Fig 21.1, the capacitor takes finite amount of time to charge to the value of voltage that is applied across its terminals. This phenomenon is generally referred to as called charging of a capacitor in RC Circuit.

When the DC Voltage applied across the Capacitor is removed, it discharges to the value of 0V almost instantly. However, the capacitor takes finite amount of time to discharge to 0V in a RC circuit. This phenomenon is generally referred to as discharging of a capacitor in RC Circuit.

RC circuits find wide applications in electronics. Understanding the charging and discharging behavior of RC circuits is fundamental to understanding electronics. This chapter deals with the concept of RC Charging and discharging when a DC supply is applied.

RC Charging Circuit

As shown in Fig 21.1 below, R1 and C1 are connected in series. The voltage at the node Vout is assumed to be 0. This is because the capacitor is initially in the uncharged state. This is done by .ic directive. The .ic directive allows one to specify initial conditions for transient analysis. Using this directive Node voltages and inductor currents can be specified. Our aim here in the analysis of RC circuits is to see the charging behavior of the circuit with respect to time. Laplace technique is used to get the expression of the charging behavior of the RC Circuit.

The expression of charging of a capacitor in RC Circuit is obtained in equation 21.9

$$V_{out} = V_i \left(1 - e^{\left(\frac{-t}{\tau} \right)} \right)$$

where $\tau = RC$

Here τ is the time constant of a RC Circuit. This is the parameter of most importance to us as it tells us the time required for the capacitor to charge to V_i.

A capacitor in RC circuit charges to 63% of V_i in 1τ seconds and it takes 5τ seconds for the capacitor to charge to approximately 100% of V_i.

RC Charging Circuit with R1=10k and C1=10μ

RC Charging Circuit with R1=10k and C1=10μ is shown in Fig 21.1 below. As per the discussion above, for this circuit $\tau = RC = 0.1$ second .This means that in this circuit the capacitor charges to 63% of V_i in 0.1 second and to approximately 100% in 0.5 seconds.

Capacitor Charging and Discharging (RC Circuit - Time Response)

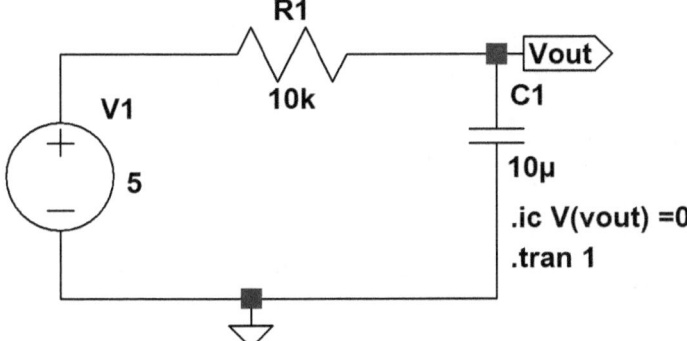

Fig 21.1: RC Charging Circuit with R1=10k and C1=10μ

RC Circuit Charging- Analysis using Laplace Transform

Applying Kirchoffs Voltage Law(KVL) to the circuit we have,

$$V_1 - I_1R - V_{out} = 0 \qquad\qquad 21.1$$

We know that for a Capacitor,

$$I = C\frac{dV}{dt} \qquad\qquad 21.2$$

Using 21.1 and 21.2 we have,

$$V_1 - V_{out} = I_1R = RC\frac{dV_{out}}{dt} \qquad\qquad 21.3$$

$$V_1 = RC\frac{dV_{out}}{dt} + V_{out} \qquad\qquad 21.4$$

Applying Laplace transform on both sides of 21.4 and keeping in knowledge that V_1 is a step input we have,

$$\frac{V_1}{s} = RCsV_{out}(s) + V_{out}(s) \qquad\qquad 21.5$$

$$\frac{V_1}{s} = (1 + RCs)V_{out}(s)$$

$$V_{out}(s) = \frac{V_1}{s(1 + RCs)}$$

Dividing the numerator and denominator by RC we have,

$$V_{out}(s) = \frac{V_1}{RC}\left(\frac{1}{s\left(s+\frac{1}{RC}\right)}\right) \qquad 21.6$$

Using partial fraction expansion technique we get,

$$V_{out}(s) = \frac{V_1}{RC}\left(\frac{\alpha}{s}+\frac{\beta}{\left(s+\frac{1}{RC}\right)}\right) \qquad 21.7$$

$$\frac{1}{s\left(s+\frac{1}{RC}\right)} = \left(\frac{\alpha}{s}+\frac{\beta}{\left(s+\frac{1}{RC}\right)}\right)$$

Solving for α and β we have,

$$\alpha s + \alpha\frac{1}{RC}+\beta s = 1$$
$$\Rightarrow \alpha + \beta = 0$$
or, $\alpha = -\beta$

also,

$$\alpha\frac{1}{RC}=1$$

$$\Rightarrow \alpha = RC \text{ and } \beta = -RC$$

$$V_{out}(s) = \frac{V_1}{RC}\left(\frac{RC}{s}-\frac{RC}{\left(s+\frac{1}{RC}\right)}\right)$$

$$V_{out}(s) = V_i\left(\frac{1}{s}-\frac{1}{\left(s+\frac{1}{RC}\right)}\right) \qquad 21.8$$

Capacitor Charging and Discharging (RC Circuit - Time Response)

Applying Inverse Laplace transform on both sides of 21.8

$$V_{out} = V_1\left(1 - e^{\left(\frac{-t}{RC}\right)}\right)$$ 21.9

Since in our present simulation $V_1 = 5V, R = 10k$ and $C = 10\mu$

$$V_{out} = 5\left(1 - e^{\left(\frac{-t}{0.1}\right)}\right)$$ 21.10

$$V_{out} = 5\left(1 - e^{\left(\frac{-t}{\tau}\right)}\right)$$ 21.11

where $\tau = RC$

The plot below in Fig 21.2 shows the level up to which the capacitor charges with respect to time.

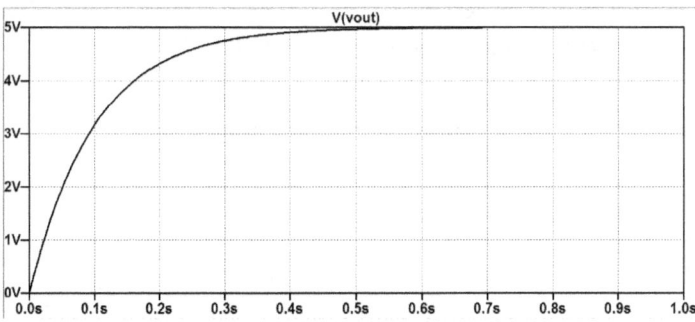

Fig 21.2: Transient analysis- RC Charging Circuit with R1=10k and C1=10μ

RC Charging Circuit with R1=100k and C1=10μ

RC Charging Circuit with R1=100k and C1=10μ is shown in Fig 21.3 below. As per the discussion above, for this circuit $\tau = RC = 1$ second .This means that in this circuit the capacitor charges to 63% of V_1 in 1 second and to approximately 100% in 5 seconds.

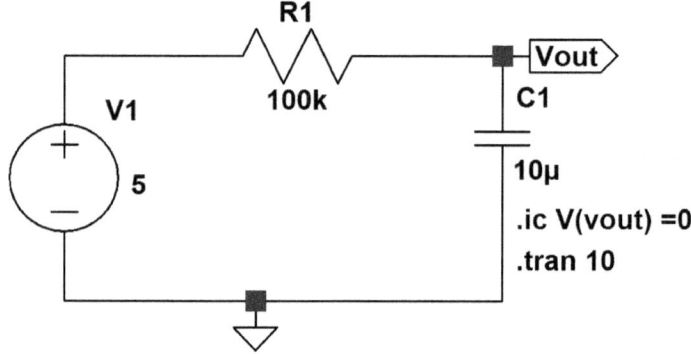

Fig 21.3: RC Charging Circuit with R1=100k and C1=10μ

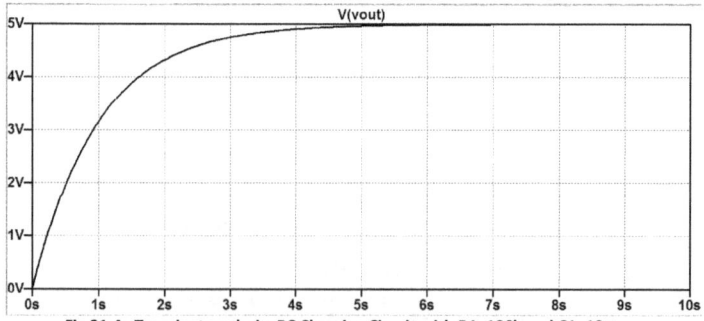

Fig 21.4: Transient analysis- RC Charging Circuit with R1=100k and C1=10μ

RC Discharging Circuit

As shown in Fig 21.5 below, R1 and C1 are connected in series but the supply voltage has been removed from the original circuit. The voltage at the node Vout is assumed to be 5V. This is because the capacitor is now assumed to be in fully charged state. This is done by .ic directive as before. Our aim here in this analysis is to see the discharging behavior of the circuit with respect to time.

RC Discharging Circuit with R1=100k and C1=10μ

RC Discharging Circuit with R1=100k and C1=10μ is shown in Fig 21.5 below. As per the discussion above, for this circuit $\tau = RC = 1$ second. This means that in this circuit the capacitor discharges to 37 % of Vout in 1 second and to approximately 100% in 5 seconds.

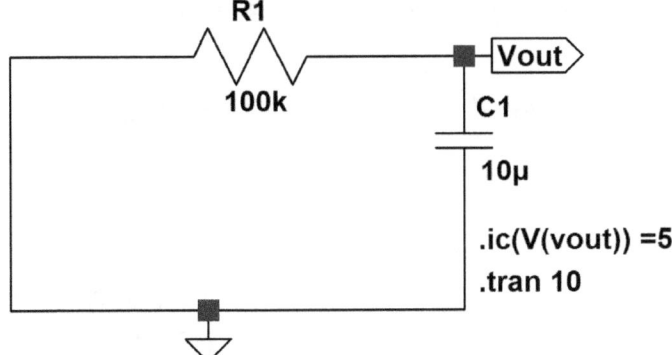

Fig 21.5: RC Discharging Circuit with R1=100k and C1=10μ

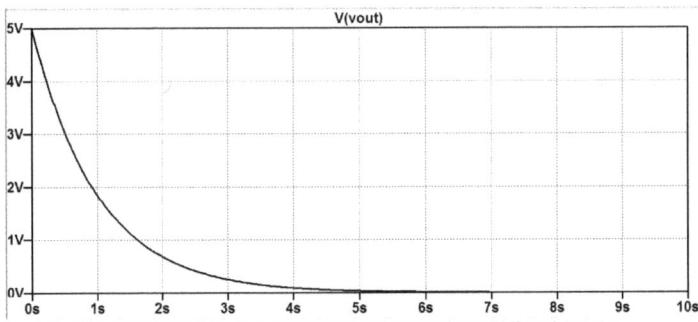

Fig 21.6: Transient analysis- RC Discharging Circuit with R1=100k and C1=10μ

Electronics Circuit SPICE Simulations with LTspice

RC Discharging Circuit with R1=5k and C1=100μ

RC Discharging Circuit with R1=5k and C1=10μ is shown in Fig 21.7 below. As per the discussion above, for this circuit $\tau = RC = 0.5\,second$. This means that in this circuit the capacitor discharges to 37 % of Vout in 0.5 second and to approximately 100% in 2.5 seconds.

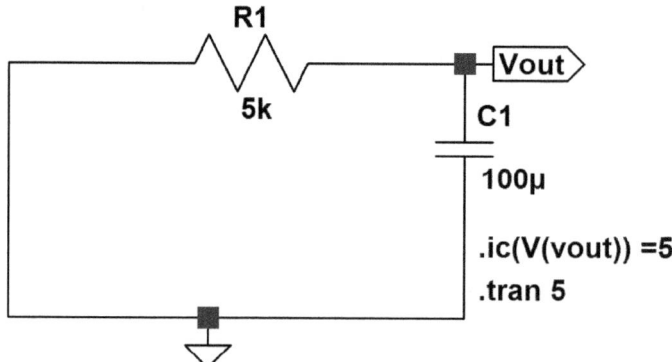

Fig 21.7: RC Discharging Circuit with R1=5k and C1=100μ

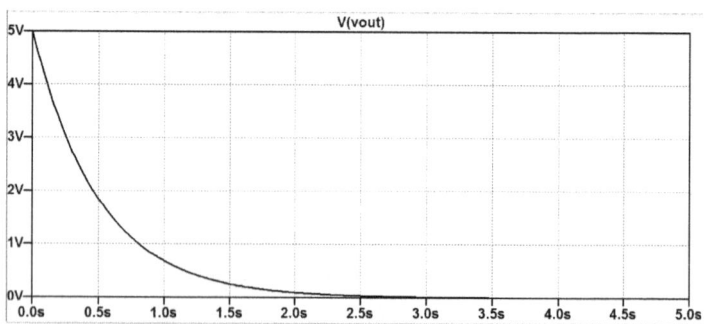

Fig 21.8: Transient analysis- RC Discharging Circuit with R1=100k and C1=10μ

****

Chapter 22   RLC Circuit - Time Response

The concept of a RLC Circuit (The case of DC Supply Voltage)

RLC Circuit: R= Resistor, L= Inductor, C=Capacitor. RLC circuits may have different configurations. Series and parallel RLC circuits are the simplest ones. This chapter handles the series configuration. Any electrical circuit does have a resistor, capacitor and an inductor built in it. The only trick, these values may differ widely. There may be circuits with high resistive value but little capacitance and inductance values or a circuit with high capacitance value and low resistance and inductance value. These values may sometimes be so small that their effect may not be clearly visible. However this chapter deals with a circuit that has typical values of resistance, capacitance and inductance that will lead to an oscillatory behavior in the time response. A very important point to note in a RLC circuit is that the resistor dissipates energy in terms of heat, while a capacitor and inductor store energy. RLC circuits find wide applications as different kinds of filters and oscillators.

RLC Circuit: With DC Power Supply

As shown in Fig 21.1 below R1, L1 and C1 are connected in series. There are different parameters like the current through the inductor or the voltage across the resistor that one can measure. But in this chapter voltage across the capacitor i.e. Vout is the focus of our study. The voltage at the node Vout is assumed to be 0V. This is because the capacitor is initially in the uncharged state. The current across the inductor I(L1) is assumed to be 0A. This is because the inductor has no current through it initially. This is done by .ic directive. The .ic directive allows one to specify initial conditions for transient analysis. Using this directive node voltages and inductor currents can be specified. Our aim here in the analysis of RLC circuits is to observe the dynamic behavior of the circuit with respect to time. Laplace technique is used to get the expression of the dynamic behavior of the RLC Circuit.

From equation 21.5 we can see that the RLC circuit can be represented by a second order differential equation. Based on different values of R, L and C, four kinds of output waveforms or solutions to this differential equation can be obtained.

1) Overdamped
2) Underdamped
3) Undamped
4) Critically damped

Again the focus of this chapter is only the Underdamped response. However other responses can also be obtain by varying the values of R, L and C.

The values of R, L and C values taken in Fig 21.1 results in an Underdamped response as given by equation 21.15.

$$V_{out} = 5 - 5e^{-500t} \cos(9987.5t) - 0.25e^{-500t} \sin(9987.5t)$$

RLC Circuit with R1=100, L1=100m and C1=100n

RLC Circuit with R1=100, L1=100m and C1=100n is shown in Fig 21.1. Transient Analysis for 10miliseconds is performed. From Fig 21.2 the Underdamped response of the circuit can be seen. Vout initially oscillates and shoots above the DC Supply (5V) and then settles down for a steady state response.

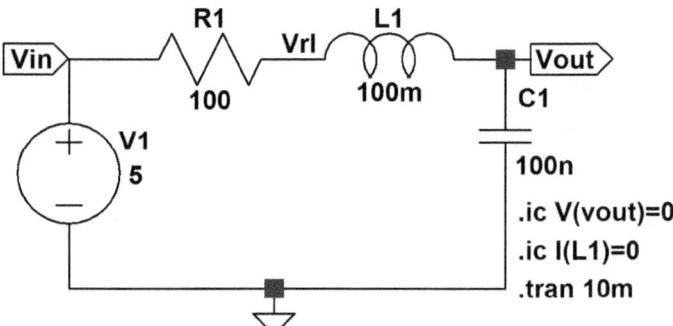

Fig 21.1: RLC Circuit with R1=100, L1=100m and C1=100n

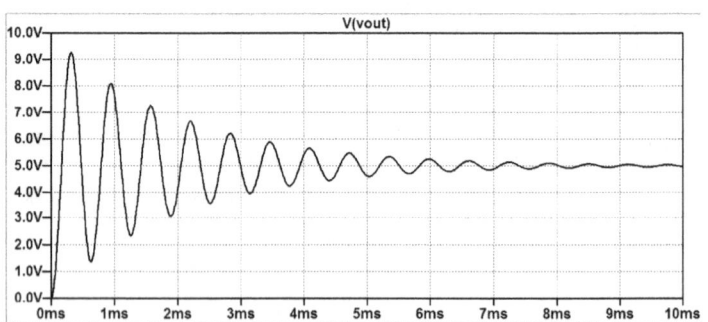

Fig 21.2: Transient analysis (Vout)- RLC Circuit with R1=100, L1=100m and C1=100n

RLC Circuit - Time Response

Fig 21.3 below shows the transient analysis of currents through R1, L1 and C1 and fig 21.4 shows the transient analysis of various voltages for the RLC Circuit shown in fig 21.1.

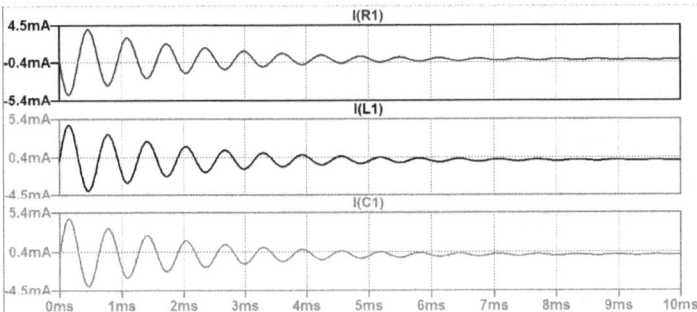

Fig 21.3: Transient analysis (current through R1, L1 and C1)- RLC Circuit with R1=100, L1=100m and C1=100n

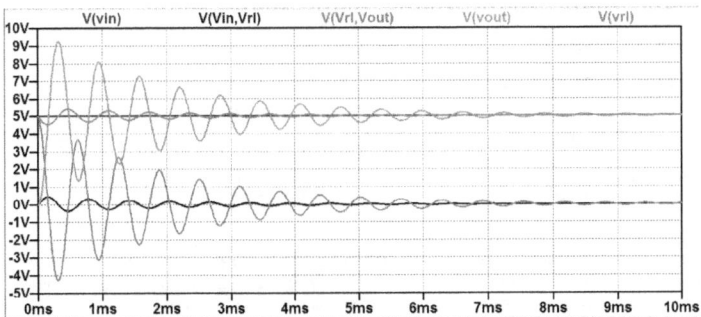

Fig 21.4: Transient analysis (Various voltages)- RLC Circuit with R1=100, L1=100m and C1=100n

RLC Circuit - Analysis using Laplace Transform

Applying Kirchoffs Voltage Law(KVL) to the circuit we have,

$$V_i - I_i R - V_L - V_{out} = 0 \qquad\qquad 21.1$$

We know that for a Capacitor,

$$I = C\frac{dV}{dt} \qquad\qquad 21.2$$

We know that for a Inductor,

$$V = L\frac{dI}{dt} \qquad\qquad 21.3$$

Using 21.1 , 21.2 and 21.3 we have,

$$V_i - V_{out} = I_i R + V_L = RC\frac{dV_{out}}{dt} + LC\frac{d^2V_{out}}{dt^2} \qquad\qquad 21.4$$

$$V_i = LC\frac{d^2V_{out}}{dt^2} + RC\frac{dV_{out}}{dt} + V_{out} \qquad\qquad 21.5$$

Applying Laplace transform on both sides of 21.5 and keeping in knowledge that V_i is a step input we have,

$$\frac{V_i}{s} = LCs^2 V_{out}(s) + RCsV_{out}(s) + V_{out}(s) \qquad\qquad 21.6$$

$$\frac{V_i}{s} = \left(1 + RCs + LCs^2\right)V_{out}(s)$$

$$V_{out}(s) = \frac{V_i}{s\left(1 + RCs + LCs^2\right)} \qquad\qquad 21.7$$

Equation 21.7 gives the transfer function of the RLC circuit in Laplace domain. In order to get the time domain solution of this transfer function, Inverse Laplace transform is taken. If there exists a direct inverse Laplace transform things are easier as Laplace transform table can be used else partial fraction technique is used to simplify the transfer function.

RLC Circuit - Time Response

$$V_{out}(s) = V_1 \left(\frac{1}{s\left(LCs^2 + RCs + 1\right)} \right) \qquad 21.8$$

Using partial fraction expansion technique we get,

$$V_{out}(s) = V_1 \left(\frac{\alpha}{s} + \frac{\beta s + \gamma}{\left(LCs^2 + RCs + 1\right)} \right) \qquad 21.9$$

$$\left(\frac{1}{s\left(LCs^2 + RCs + 1\right)} \right) = \left(\frac{\alpha}{s} + \frac{\beta s + \gamma}{\left(LCs^2 + RCs + 1\right)} \right)$$

Solving for α, β and γ we have,

$$\alpha LCs^2 + \alpha RCs + \alpha + \beta s^2 + \gamma s = 1$$

By comparing the coefficients on both sides we have,

$$\alpha = 1$$

$$\alpha LC + \beta = 0$$

and $\alpha RC + \gamma = 0$

$$\Rightarrow \alpha = 1, \ \beta = -LC \ \text{and} \ \gamma = -RC$$

$$V_{out}(s) = V_1 \left(\frac{1}{s} - \frac{LCs + RC}{\left(LCs^2 + RCs + 1\right)} \right) \qquad 21.10$$

Simplifying the second term on the right hand side,

$$V_{out}(s) = V_1 \left(\frac{1}{s} \right) - V_1 \left(\frac{s + \dfrac{R}{L}}{\left(s^2 + \dfrac{R}{L}s + \dfrac{1}{LC}\right)} \right) \qquad 21.11$$

In our case, $V_1 = 5$, $R = 100$, $L = 100m$, $C = 100n$

$$V_{out}(s) = 5 \left(\frac{1}{s} - \frac{(s + 1000)}{\left(s^2 + 1000s + 100000000\right)} \right) \qquad 21.12$$

Electronics Circuit SPICE Simulations with LTspice

$$V_{out}(s) = 5\left(\frac{1}{s} - \frac{(s+1000)}{\left(s^2 + 1000s + 250000 + 99750000\right)}\right)$$

$$V_{out}(s) = 5\left(\frac{1}{s} - \frac{(s+1000)}{\left((s+500)^2 + 99750000\right)}\right)$$

$$V_{out}(s) = 5\left(\frac{1}{s} - \frac{(s+1000)}{\left((s+500)^2 + (9987.5)^2\right)}\right) \qquad 21.13$$

$$V_{out}(s) = 5\left(\frac{1}{s} - \frac{(s+500)}{\left((s+500)^2 + (9987.5)^2\right)}\right) - 5\left(\frac{500}{\left((s+500)^2 + (9987.5)^2\right)}\right)$$

$$V_{out}(s) = 5\left(\frac{1}{s}\right) - 5\left(\frac{(s+500)}{\left((s+500)^2 + (9987.5)^2\right)}\right) - 5\left(\left(\frac{500}{9987.5}\right)\frac{9987.5}{\left((s+500)^2 + (9987.5)^2\right)}\right)$$

$$V_{out}(s) = 5\left(\frac{1}{s}\right) - 5\left(\frac{(s+500)}{\left((s+500)^2 + (9987.5)^2\right)}\right) - 5\left(\left(\frac{500}{9987.5}\right)\frac{9987.5}{\left((s+500)^2 + (9987.5)^2\right)}\right)$$

$$V_{out}(s) = 5\left(\frac{1}{s}\right) - 5\left(\frac{(s+500)}{\left((s+500)^2 + (9987.5)^2\right)}\right) - 0.25\left(\frac{9987.5}{\left((s+500)^2 + (9987.5)^2\right)}\right) \qquad 21.14$$

$$V_{out} = 5 - 5e^{-500t}\cos(9987.5t) - 0.25e^{-500t}\sin(9987.5t) \qquad 21.15$$

Typical solution:

let $t = 0.30\,msec$

$V_{out} = 5 - 5e^{-500 \times 0.30m}\cos(9987.5 \times 0.30m) - 0.25e^{-500 \times 0.30m}\sin(9987.5 \times 0.30m)$

$V_{out} = 5 - 5e^{-0.15}\cos(2.99) - 0.25e^{-0.15}\sin(2.99)$

$V_{out} = 5 - 5(0.86)(-0.988) - 0.25(0.866)(0.15)$

$V_{out} = 5 + 4.248 - 0.032 = 9.216$

Thus $V_{out} = 9.216$ at $0.30\,msec$.

RLC Circuit - Time Response

RLC Circuit: With DC Power Supply removed

As shown in Fig 21.5 below, R1, L1 and C1 are connected in series but with the DC Power supply removed. Again Vout is the focus of our study. The voltage at the node Vout is assumed to be 5V. This is because the capacitor is now initially in the charged state. The current across the inductor I(L1) is assumed to be 0.4mA. Transient Analysis for 10miliseconds is performed. From Fig 21.6 the dying response of the circuit can be seen. Vout initially is at DC Supply (5V) and then gradually settles down for a steady state response. Fig 21.7 below shows the transient analysis of currents through R1, L1 and C1 and fig 21.8 shows the transient analysis of various voltages for the RLC Circuit shown in fig 21.5.

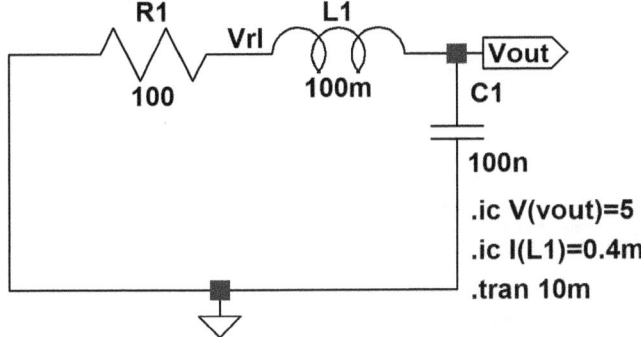

Fig 21.5: RLC Circuit with R1=100, L1=100m and C1=100n (With DC Power Supply removed)

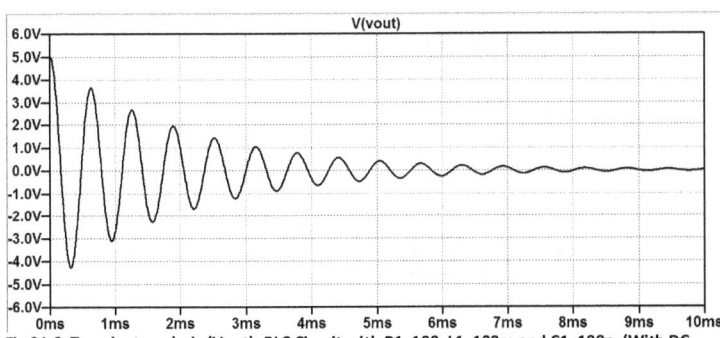

Fig 21.6: Transient analysis (Vout)- RLC Circuit with R1=100, L1=100m and C1=100n (With DC Power Supply removed)

Electronics Circuit SPICE Simulations with LTspice

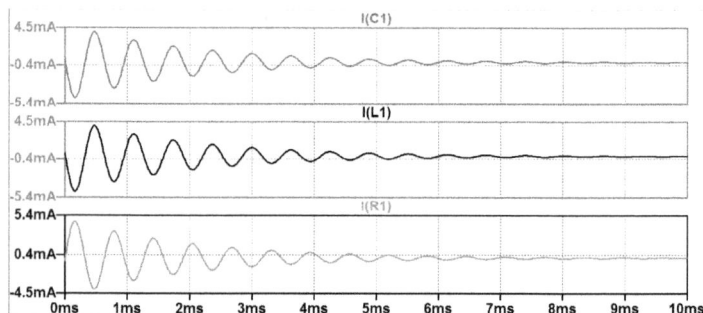

Fig 21.7: Transient analysis (current through R1, L1 and C1)- RLC Circuit with R1=100, L1=100m and C1=100n (With DC Power Supply removed)

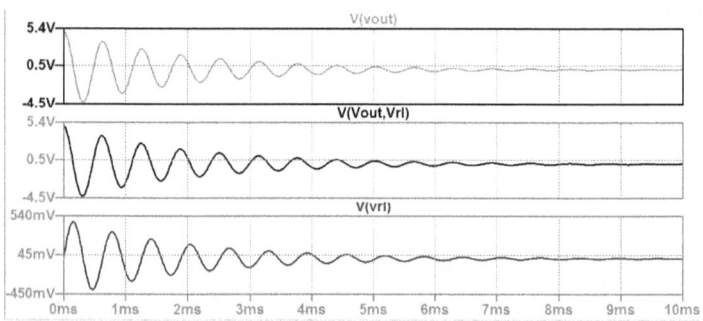

Fig 21.8: Transient analysis (Various voltages)- RLC Circuit with R1=100, L1=100m and C1=100n (With DC Power Supply removed)

****

References

[1] Engelhardt, Mike. "SPICE Differentiation." *ELECTRONICS WORLD* 121, no. 1946 (2015): 16-21.

[2] Robert, Boylestad, and Nashelsky Louis. "Electronic devices and circuit theory." (1995).

[3] Mehta, V. K., and Rohit Mehta. "Principle of Electronics." (2011).

[4] Choudhury, D. Roy, and Shail Jain. *Linear integrated circuits*. John Wiley & Sons, Inc., 1991.

[5] Malvino, Albert, and David Bates. *Electronic Principles*. McGraw-Hill, Inc., 2006.

[6] Chen, Wai-Kai, ed. Analog circuits and devices. CRC Press, 2003.

[7] Mac Elwyn Van Valkenburg. *Network analysis*. Prentice Hall, 1955.

[8] LTspice Help Topics available inbuilt with LTspice.

[9] http://ltspice.linear.com/software/LTspiceGettingStartedGuide.pdf as on 30th July 2015

[10] http://cds.linear.com/docs/en/software-and-simulation/LTspiceIV_flyer.pdf as on 30th July 2015

About the Authors

Amit Kumar Singh Obtained his M.Tech degree in VLSI Systems and Technology from School of Engineering, Shiv Nadar University, Dadri in 2014. He obtained his B.Tech degree in Electronics and Telecommunication Engineering from Faculty of Engineering and Technology, SRM University, Kattankulathur in 2010. He is actively involved in research and teaching activities. His areas of interest include Digital Signal Processing and VLSI.

Rohit Singh Obtained his M.Tech degree in VLSI Systems and Technology from School of Engineering, Shiv Nadar University, Dadri in 2014. He obtained his B.Tech degree in Electronics and Communication Engineering from Babu Banarsi Das Institute of Technology Ghaziabad, affiliated to UPTU in 2012. He is actively involved in research activities. His current research area is fabrication and modeling of transistors.

Electronics Circuit SPICE Simulations with LTspice

A Schematic Based Approach

First Edition

Amit Kumar Singh

Rohit Singh

The book covers the requirements of a laboratory course in SPICE simulations at an introductory level. It can be used an aid to practical understanding in any undergraduate engineering course of Analog electronics. The book can also be used as an aid to any standard text on Analog Electronics.

Salient Features:

- *Step by step simulation procedure is presented*

- *Experiments are clearly illustrated.*

- *Brief theory on each topic for understanding is presented.*